CHINA'S
SATELLITE / PARTIES

Studies of the East Asian Institute—Columbia University

The East Asian Institute of Columbia University

The East Asian Institute is Columbia University's center for research, publication, and teaching on modern East Asia. The Studies of the East Asian Institute were inaugurated in 1962 to bring to a wider public the results of significant new research on Japan, China, and Korea.

CHINA'S
SATELLITE PARTIES

James D. Seymour

An East Gate Book

M. E. Sharpe, Inc.
Armonk, New York/London

East Gate Books are edited by Douglas Merwin
120 Buena Vista Drive, White Plains, New York 10603

Copyright © 1987 by M. E. Sharpe, Inc.

All rights reserved. No part of this book may be reproduced in any form without written permission from the publisher, M. E. Sharpe, Inc., 80 Business Park Drive, Armonk, New York 10504

Available in the United Kingdom and Europe from M. E. Sharpe, Publishers, 3 Henrietta Street, London WC2E 8LU.

Library of Congress Cataloging-in-Publication Data

Seymour, James D.
    China's satellite parties.

    Bibliography: p.
    Includes index.
    1. Political parties—China. 2. China—
Politics and government—1949-   . I. Title.
JQ1519.A45S49   1987   324.251'009   86-31343
ISBN 0-87332-412-9

Printed in the United States of America

Contents

Preface		*vii*
Abbreviations		*xi*
1	Communism and United Fronts	3
2	China's Middle Parties and the Revolution	12
3	The DPGs and Their Role, 1950-1955	25
4	The Year of Transitions, 1956	39
5	The DPGs and "Hundred Flowers," 1957	46
6	The DPGs in Eclipse, 1958-1978	63
7	The DPGs Today	68
8	Conclusion	85
Appendix 1: The Democratic League's 1957 Academic Program		93
Appendix 2: Emergency Democratic League Conference, 1957		95
Appendix 3: A DPG Constitution: The Revolutionary Guomindang (1983)		98
Notes		109
Bibliography		135
Index		145
About the Author		151

Preface

In the People's Republic of China there exist a number of anomalous organizations known as *minzhu dang-pai*. This term literally means "democratic parties and groups," an English rendering that, whether or not appropriate, will be used in this book.

The groups were founded between the late 1920s and the early 1940s. They sought to be a middle force and a liberalizing influence, but they had no hope of competing for power against China's two great party-armies. Incorporated into the Communist order after 1949, they valiantly sought to be heard in 1957 but were then suppressed. After nearly dying out during the Cultural Revolution, the DPGs were resuscitated around 1980, and it is thus appropriate to reexamine the subject.

Much has changed since the 1950s, and the reincarnated democratic parties are hardly recognizable versions of their earlier selves. To understand the *new* DPGs, it was necessary to go to China and examine them firsthand. This I did during the winter of 1984-85. My intention was to ferret out subjects as randomly as possible and learn what DPG membership meant to them. Of course, in China no sampling can really take place randomly, but at least the interviewees were not preselected, and the sessions were not audited by representatives of officialdom. Actually, the research was not sanctioned by the authorities, and indeed I was unable to arrange for an institution to serve as a base or "unit." Nonetheless, no further obstacles were placed in my way. No help was sought or received from government or Communist Party authorities.

Altogether more than a dozen DPG members from almost as many branches were interviewed. This sample is so small that the findings may be termed anecdotal. Nonetheless, if one allows for certain biases in the sample, the results ought to be meaningful. The main problem is the likelihood that the typical

DPG member would be unwilling to be interviewed by a foreigner, especially without authorization; he or she may be a timid follower, more in the habit of listening than talking freely, and unwilling to take chances. History has taught most Chinese that the key to survival is avoiding political risks. In short, my interviewees were doubtless more dynamic than is normal.

The following are the six groups that are the subject of this book. Aside from the two- or three-letter abbreviation, most of the groups are given shortened names (in parentheses) by which they will commonly be referred in the text.

APD   China Association for Promoting Democracy (Zhongguo minzhu cujin hui). Primary constituency: school teachers. Size (1984): 15,000.
DL   China Democratic League (the League) (Zhongguo minzhu tongmeng ["min-meng"]). Primary constituency: intellectuals in general. Size (1956): 30,000; (1984): 50,000.
September Third Study Society (Jiusan) (Jiu-san xue-she ["Jiusan"]). Primary constituency: higher intellectuals. Size (1984): 12,000.
NCA   China National Construction Association (Construction) (Zhongguo minzhu jianguo hui ["min-jian"]). Primary constituency: businessmen. Size (1956): 25,000; (1984): 25,000.
PW   Chinese Peasants' and Workers' Democratic Party. (Zhongguo nong-gong minzhu dang). Primary constituency: health professionals.
RG   Guomindang Revolutionary Committee (Revolutionary Guomindang) (Guomindang geming weiyuanhui ["min-ge"]). Primary constituency: former Nationalist Party members. Size (1956): 16,000; (1984): 20,000.

The total DPG population is now said to have been 20,000 in 1949. Membership climbed to 100,000 in 1956 (and to a larger but unknown figure by 1957) but dropped to 65,000 during the Cultural Revolution. After 1978 the DPGs grew, reaching 160,000 by early 1986.

The two groups not covered in this work (but normally considered DPGs) are the China Public Interest Party (Zhongguo zhigong dang) and the Taiwan Democratic Self-Government League (Taiwan minzhu zizhi tongmeng), or "Tai-meng" for short. The Public Interest Party is composed of 2,300 returned overseas Chinese (1983 figure). The Taiwan League, whose purpose is to keep alive the notion that native Taiwanese wish Taiwan to be part of the People's Republic of China, is doubtless even smaller.

Even by the standards of academia, this small book took a long time to complete. The first phase of the research began in 1959, culminating in a Master's thesis for Columbia University's Department of Public Law and Government. During the 1960s the democratic parties and groups declined, and other aspects of Chinese politics demanded the attention of China watchers. I did not follow up on this early work, and the thesis was not published. But my Master's essay appears to have stood the test of time, and the story of the exciting DPG developments of the 1950s has remained largely untold. Thus, the first half of this book is a slightly revised version of that essay. The remainder of the book is based on recent research and interviews.

There are numerous people whom I wish to thank for the help they have provided. First, very belated thanks are due to O. Edmund Clubb, Edward Friedman, and Lyman P. Van Slyke, who provided much useful advice in the early 1960s. Next, there are my colleagues at Columbia University's East Asian Institute (Myron L. Cohen, Andrew Nathan, Andrew Walder, and others) who offered many useful suggestions when I presented chapter 7 at a seminar in 1985. Likewise, the productiveness of my several brief periods as scholar-in-residence at Hong Kong's Universities Service Centre--most recently the winter of 1984-85--owes in large measure to John Dolfin and Chan Hing Ho. Alan Pauw kindly let me rely on his unpublished paper on the DPGs during 1960s and 1970s, which was a great help to me in writing chapter 6. Anita Chan read an early version of the entire book and made countless valuable suggestions. David Eisner, Kenneth Lieberthal, Richard Madsen, Roderick MacFarquhar, Doug Merwin, Richard Sorich, and Jonathan Unger also supplied information and advice.

But my greatest debt is owed to those in China who either agreed to be interviewed or arranged interviews. Most of them were promised anonymity, and it would not be appropriate to name even those who were not. I hope that the result justifies the risk that they all ran in becoming involved in this project. I am extremely grateful to them. It cannot be overemphasized, however, that errors of fact and interpretation are mine alone.

In spelling Chinese names, I have used the pinyin system for the most part. When quoting from English-language sources, I have changed spellings to pinyin. When material has been adapted from my 1961 Master's essay, retransliteration from the Wade-Giles system has involved some guesswork. Any resulting misspellings are regretted.

Abbreviations

APD	China Association for Promoting Democracy
CPPCC	Chinese People's Political Consultative Conference
CP	Communist Party
DL	China Democratic League
DPG	democratic party or group (*minzhu dang pai*)
NCA	China National Construction Association
NPC	National People's Congress
PPCC	People's Political Consultative Conference (at levels below the national CPPCC)
PRC	People's Republic of China
PW	Chinese Peasants' and Workers' Democratic Party
RG	Revolutionary Guomindang, or Guomindang Revolutionary Committee
UFD	United Front (Work) Department

1
Communism and United Fronts

Communist revolutions have generally been products of intellectuals rather than proletarians. George Konrád and Ivan Szeléyi have suggested that so-called proletarian revolutions in Russia and Eastern Europe were essentially power-grabs by intellectuals.[1] Although it would be misleading to characterize the Chinese revolution this way, it is true that *some* intellectuals have ended up in privileged positions. These include the members of the minor political parties, organizations which have European counterparts. Indeed, notwithstanding the difficulty one has in finding sanction for such a phenomenon in the writings of Marx and Lenin,[2] many Communist countries sport "non-Communist" parties and "coalition" or united front governments.

Non-Communist Parties in Communist States

In Soviet Russia itself, the multiparty system was done away with a few years after the 1917 revolution. Unlike the non-Communist parties that later would exist in Eastern Europe and China, the most important party in Russia coexisting with the Bolsheviks consisted of a genuine, Marxist opposition. For four years, the Mensheviks endeavored, by constitutional means, to oust the Bolshevik Party. Their presence was tolerated, but their activities were interfered with and individuals were harassed. Their economic program was a liberal one (similar to the Bolsheviks' later New Economic Policy), and during 1919 and 1920 it earned them a growing influence in the trade unions. The end of the Menshevik Party came early in 1921 with a series of arrests. There was nothing legal about its destruction, which came about (in the words of one Western scholar) "by force and by fraud."[3]

The Eastern European Communist parties that came to power after World War II usually did not find it expedient to terminate their old rivals so cavalierly. Typically, the non-Communist parties had important histories and large public followings. Thus, the new governments often found it useful to permit their continued existence--at least in name. In at least one of these countries (Czechoslovakia) a multiparty system was established to which China's would bear a striking resemblance, with the non-Communist parties rendering real services to the ruling elite. In most East European countries, however, non-Communist parties had been so tainted by the flavor of genuine opposition that it was in the interest of the Communists to reduce their strength in every feasible way, which in some cases (Hungary, Rumania, and Bulgaria) did mean extinction.[4]

But to this day, other Eastern European countries have satellite parties, including Poland,[5] Czechoslovakia, and East Germany. The latter case presents an interesting parallel with China, because the parties are supposed to be counterparts of their West German namesakes. The Christian-Democratic Union and the Liberal Democratic Party were originally given quotas in the Volkskongress, with Soviet authorities removing individuals who did not cooperate with the local Communist leaders. Some ministries were headed by non-Communists, but they nearly always proved to be staunch supporters of the regime. The parties themselves simply perpetuated empty existences. The authorities would not allow them to dissolve, but only by infiltrating their organizations disguised as Communists was it able to prevent this from happening.[6] These parties have a close analogue in China's Revolutionary Guomindang.

It is in the case of Czechoslovakia, however, that China's multiparty system finds its most genuine parallel. On the subject of Czech non-Communist parties we are fortunate in having an early study by Edward Taborsky, who was himself an official of the government prior to the 1948 coup. In Czechoslovakia, an avowed function of the non-Communist parties is to wait on the Communist Party, "to hail its leadership" and act as its "handymen" in the realization of Marxist-Leninist goals."[7] The ruling Party reorganized the old parties along the lines of "democratic centralism." Only trusted individuals were placed in leadership positions within them. A party functionary was responsible not only to his immediate superiors but also to the Communist Party organ on his own level. These and other features closely parallel the Chinese pattern, especially during the 1950s--the small size of the parties, their Communist-line newspapers, the parties' roles in indoctrination (through seminars, etc.), and the fact that non-Communist candidates, picked by the Communists and running under a "National Front" label, stand for election to political office.

But while the Chinese and Czech situations appear to have mirrored each other, it is not correct to conclude that one is a copy of the other. The decisions of the Czech and Chinese Communists to maintain non-Communist parties were probably made independently of each other. The Czech Communists were the first to act in this direction, in 1948, but the Chinese had already committed themselves to a multiparty state. If there was ever any conscious imitation (and there is no evidence of this), it could have not have involved more than the question of the use to which the parties ultimately were put.

Probably the objective conditions in both countries called for the maintenance of non-Communist parties, but it is surprising how dissimilar those conditions were in many crucial respects. For one thing, until the completion of industrialization China was seen as needing a bourgeoisie, and therefore bourgeois-democratic parties. In Czechoslovakia this factor was less emphasized, for the country was as industrialized in 1948 as any country has ever been at the time of "liberation." Another distinction lies in the earlier significance of the non-Communist parties. In Czechoslovakia they had been very important, with a combined strength about equal to that of the Communists prior to the latter's accession to power. In China the small middle parties had never had any real impact on the political life of the nation. They had no armies, and they generally carried the flavor of Western liberalism so alien to the Chinese people. Finally, unlike China, Czechoslovakia had experienced Western-style democratic institutions. These could be discarded more gracefully if the fiction of a multiparty system was maintained. In the light of these three features that distinguished the Czech and Chinese situations, then, it is evident that the respective Communist parties' decisions to perpetuate the lives of their old domestic rivals were not dictated by the same considerations. What the two movements did have in common was the heritage of experiences with united fronts and coalition governments.

The Chinese Communists' Pre-1949
United Front Strategy

The emergence of the united front strategy in world communism involved the tacit admission that proletarian revolutions were less likely to occur in industrialized countries with large urban working classes than in underdeveloped regions, previously exploited by foreign capitalists, where proletariats were virtually nonexistent. At first the concept was relatively simple: A weak communist party should ally with all progressive and anti-imperialist forces. The strategy that was ultimately to help carry Mao Zedong and his followers to power, however, was more subtle and discriminating than this.

Communist parties were first required to engage in united fronts at the Second Comintern Congress of 1920. Under the direction of Lenin, this congress decided that affiliates in colonial and semicolonial countries should cooperate with any anti-imperialist groups. The following year the Communist Party of China (CP) was organized. China was seen as a semicolony, so it was appropriate for the Communist Party there to ally with the Guomindang, which it did in 1923. The alliance was an uneasy one, and it disintegrated in 1927.

The 1930s saw the ascendancy to power within the CP of Mao Zedong. Before this, the experiments in interclass and interparty alliances had been erratic (due in part to disagreements within the Communist Party) and generally unsatisfactory. Mao had learned that it was necessary to avoid both the pre-1927 "rightist-opportunist" mistake of "all alliance and no struggle" and the post-1927 "leftist-opportunist" mistake of "all struggle and no alliance."[8] But the task of working out a new compromise was complicated by the dual purpose that any cooperative arrangement had to serve. An alliance that did not further the cause of the revolution would not satisfy the Communists. But the immediate issue was how to resist effectively the "imperialists"--now meaning the Japanese.

How the Communists in 1937 resolved the conflicting requirements of the situation is a complicated story, and it is not necessary to be concerned with all the details. Briefly, after the Japanese struck in Manchuria in 1931, the Communists saw a new necessity to cooperate with other groups. This need was expressed the following year, but it was not clear whether the Guomindang was included in their plans. The final consolidation of Mao Zedong's authority over the Party in 1935 marked a further step away from "closed-door sectarianism." In 1937, after the outbreak of the Sino-Japanese War, an agreement was worked out with the Guomindang, and for a time the two parties cooperated in a meaningful way. But relations showed signs of strain in 1938, and in the following years there were several instances of open hostilities between them. Nominal policy remained the same, however, and the Communists did cooperate with non-Communists in administering areas under Communist control. The year 1940 saw the introduction of the "three-thirds" system, according to which Communists, Guomindang people, and others ("middle groups" and non-Party people) were represented in government agencies in equal proportions. What is significant for our purposes is the favorable impression the system made on the middle groups.

Although policies during these years were dictated primarily by the threat of Japanese imperialism, it was considerations relating to the revolution that underlay the long-range thinking of the Communists regarding interclass cooperation. On this subject we learn more from the theoretical writings of Mao Zedong than from studying the expedient policies actually adhered to in the 1930s and early 1940s. Mao's thinking on class outlooks can be traced back to his earliest writings, particularly his "Analysis of the Classes of Chinese Society" (1926).[9]

But his theory of classes did not become crystallized until 1939, with the essay "The Chinese Revolution and the Chinese Society." In this essay, Mao outlined the class basis of the anticipated revolution. The lower urban classes and 90 percent of the peasantry were potential allies of the revolution, and the landlord class was definitely opposed to it. Between these two extremes lay the bourgeoisie, a more complicated class than either of these. The "big" (compradore) bourgeoisie was allied with imperialism and had no sympathy for the revolution. Occasionally, however, the imperialist powers would fight among themselves. Thus, if the united front was fighting some of those powers, part of the big bourgeoisie might find itself on the side of the revolution. Cooperation with the "national bourgeoisie" (patriotic businessmen) was feasible. This stratum sought China's industrial development and economic integrity, and it could be counted on to oppose imperialism, and perhaps to tolerate revolution. Of course, the national bourgeoisie was dual in nature. "It lacks the courage to oppose imperialism and feudalism thoroughly because it is economically and politically flabby and still has economic ties with imperialism and feudalism. This emerges clearly when the people's revolutionary forces grow powerful."[10] But there was still hope that the national bourgeoisie might side with the revolution. So far it had been a "comparatively good ally," and it was necessary to adopt a "cautious policy."

The petty bourgeoisie[11] also escaped neat categorization. This group, consisting of intellectuals, small merchants, handicraftsmen, and professionals, was generally on the side of the revolution. Nevertheless, only under the leadership of the proletariat could it be effective. Because of their backgrounds, Mao considered most of the intellectuals to be petty bourgeois. He realized that he needed the services of the intellectuals, but, like Lenin and Stalin, he did not altogether trust them.

> The intellectuals often tend to be subjective and individualistic, impractical in their thinking and irresolute in action until they have thrown themselves heart and soul into mass revolutionary struggles, or made up their minds to serve the interests of the masses and become one with them. Hence, though the mass of revolutionary intellectuals in China can play a vanguard role or serve as a link with the masses, not all of them will remain revolutionaries to the end. Some will drop out of the revolutionary ranks at critical moments and become passive, while a few may even become enemies of the revolution. The intellectuals can overcome their shortcomings only in mass struggles over a long period.[12]

Likewise, the other sectors of the petty bourgeoisie--the small merchants, handicraftsmen, and professionals--might find themselves on either side. It was necessary to pay attention to carrying out revolutionary propaganda and organizational work among such groups. Although Mao did not indicate it at this time, one day the democratic parties would be doing precisely that.

A month after this essay was written, Mao produced "On the New Democracy,"[13] in which he outlined the salient features of the socialist state he envisaged. Mao saw three basic types of states: republics under bourgeois dictatorships, republics under the dictatorship of the proletariat, and republics under the joint dictatorship of the various revolutionary classes.

China was to be the latter. This was the only type of state suited to the needs of a nation emerging from colonial or semi-colonial status. As the months went by, Mao hinted at a conciliatory attitude toward even nonrevolutionary elements; only with the most flagrant anti-Communists would he have no truck.

In 1941, further light was shed on the form the united front would take. In a speech to the Shaanxi-Gansu-Ningxia border assembly, Mao admitted that his comrades and outsiders had not always worked smoothly together, and he urged that the situation be improved. He condemned sectarianism and "closed-doorism" and insisted on the need for Communists to cooperate with others.

> We are not a small opinionated sect and we must learn how to open our doors and cooperate democratically and consult with non-Party people. Perhaps even now there are Communists who may say, "If it is necessary to cooperate with others, then leave me out." But I am sure there are very few. . . . We still have many failings. We are not afraid to admit them and are determined to get rid of them. We shall do so by strengthening education within the Party and by cooperating democratically with non-Party people. It is only by subjecting our failings to such cross-fire, both from within and from without, that we can remedy them and really set the affairs of state to rights.[14]

This statement would be quoted frequently in later years. It implied that the united front in government was to be something more than a propaganda device. It was supposed to improve the quality of administration, and even the Party itself.

Implicit in Mao's writings, and to some extent manifested in the conduct of the Chinese Communist Party before 1945, was a new kind of united front. Less emphasis was placed on its anti-imperialist underpinnings. Thus, there was less need to cooperate with the reactionary Guomindang--but Mao seemed determined not to commit the post-1927 mistake of "struggle" against everybody. The new solution was to cooperate with all groups who were potential allies of the *revolution*. This the Guomindang was not, though of course it was desirable to avoid a formal break with that party until the defeat of Japan. So the years preceding 1945 saw a steady decline in Communist-Nationalist relations, and at the same time considerable genuine cooperation between the Communists and left-wing bourgeois elements--though the latter were in a subservient position. By such tactics the Communists were able to neutralize most of China's bourgeoisie and even to count much of the class on its side when the final showdown with the Guomindang came. At a time when middle groups had been treated shabbily by the Nationalists, they were well received by the Communists. One Westerner who traveled in China reported: "I have met several members of the Democratic League in Yan'an who served in responsible government positions and gave . . . favorable opinions about the attitudes of the Communists toward the small democratic parties."[15] The nominally coalition government at Yan'an, containing left-wing bourgeois elements but led by the Communists, was essentially the post-1949 united front in its germinal form.

In April 1945, with the end of the war in sight, Mao made an important statement in "On Coalition Government."[16] Now making an even bigger play for the support of China's various middle-of-the-road parties, he announced that the New Democracy would be not only multiclass, but also multiparty. The difference between his program and Marxist-Leninist orthodoxy was frankly admitted, and it was made explicit that no longer would the united front exist for the purpose of resisting imperialism. Instead, it must serve Mao's greater purposes--the revolution and the socialist construction.

In a sense, Mao's theory resulted in a bisected bourgeoisie. Much of the class was seen as salvageable for the revolutionary cause. The remainder was counterrevolutionary and had to be fought. Mao did not draw a sharp line leaving every segment of the class to the right or left of it. Cooperative elements were promised that they, as a class, would play a political role in the New Democracy, and uncooperative elements were warned that they would be excluded. Fence sitters could anticipate a struggle to win them over. This formulation reflected the lessons of the past, which demonstrated the undesirability of being either too inclusive or too exclusive. It was also indicated that considerable thought had been given to the nature of the future Communist state. The realization that the support of at least part of the bourgeoisie was essential if that state were to succeed accounts in large measure for Mao's insistence that a modified united front policy be adhered to in the 1940s.

The Attitude of Moscow

Theoreticians in the Soviet Union apparently did not know what to make of Mao's pronouncements. In attempting to orchestrate revolutions in Asian countries, Moscow had been operating on the assumption that the entire bourgeoisie was to be cooperated with until, and only until, the proletariat was strong enough to carry on without it. Andrei Zhdanov, in his report to the first meeting of the Cominform in September 1947, was not clear as to whether or not he accepted the Maoist innovations.[17] Georgy Zhukov did accept them two months later in an article in the authoritative *Bol'shevik*.[18] But in the same month Moscow heavily edited a speech by Mao in which he said that Communist movements elsewhere should unite with "good" elements within the bourgeoisie. It was made to *appear* that he had taken a proper "leftist" view--that at this time Communists should operate independently of the bourgeoisie.[19] Only in mid-1949, with the publication in *Pravda* of an article by China's number two Communist,[20] did Moscow fully accept this aspect of Mao's actual position.

Now the only difference between the Soviet and Chinese Communist lines concerned the *duration* of the multiclass state. The Soviet view, adhered to well into the 1950s, held that the dictatorship of the proletariat must be realized when the state approached the period of the transition to socialism, whereas according to the Chinese it might wait. The difficulty was resolved only in 1956 at the Twentieth Congress of the Soviet Communist Party. It was done largely by emptying of meaning the term "dictatorship of the proletariat."[21]

2
China's Middle Parties and the Revolution

Between the revolutions of 1911 and 1949 various elements in Chinese society sought a way to adapt Western liberal political institutions to their nation. There is little in China's long history, however, to suggest that competition between political groups can be constructive. Political thinkers of the past used to emphasize the desirability of harmony; lack of harmony meant the temporary failure of the Confucian political system. The latter was not an altogether undemocratic system (the ruler had obligations to his people, and when he failed he or his dynasty could be replaced), but it was autocratic in theory, with the decision-making power at least nominally concentrated in the hands of one man--the emperor. The basic precepts of the state had been in place for two thousand years, and until modern times they were rarely questioned. It was up to the ruler to apply them to day-to-day problems. In this process, the scholar-bureaucrats could advise, and indeed had much influence, but the people at large almost always remained passive.

It is not surprising, then, that the ideas of Western liberalism were difficult for Chinese to appreciate. While such notions excited many intellectuals, others associated them with Western imperialism, which, rightly or wrongly, became a scapegoat for a people frustrated in their efforts to live the way twentieth-century human beings were supposed to be able to live. So the two political parties that did manage to rouse the political passions of the Chinese were unlike any that have existed in democratic countries. The Guomindang and the Communist Party both stood for the principle of political absolutism, having been organized according to a pattern imported from the Soviet Union--Lenin's "democratic centralism." They were not designed to respond to the yearnings of the masses so much as to organize mass support for predetermined programs.

Background of the Middle Parties

In addition to the two major parties there were any number of smaller groups offering a wide variety of programs. Some, like Huang Yanpei's Vocational Educational Group (founded in 1917), were hardly designed to serve political purposes at all. Huang discovered, however, that to implement his plans for education he needed political (as well as financial) support.[1] The China Youth Party, founded in 1918, was more frankly political. At first spurred in a liberal direction by China's postwar treatment at Versailles, after the leftists parted from its ranks to join the Communist Party it became increasingly identified with conservatism and warlord intrigue.[2]

In 1927 a small number of defectors from the Guomindang set up what was to be commonly called the Third Party.[3] As their leader they recognized Deng Yanda, who until 1930 was in self-imposed exile in Europe. From there he encouraged the movement, and he may have set up an overseas headquarters in Berlin. A few months after his return to China, Deng was executed by the Nationalists.[4] Demoralized, the Third Party split into factions. During the Sino-Japanese War it was revived as a leftist splinter group by some of the early leaders (among them, Zhang Bojun), and in 1947 its name was changed to the Chinese Peasants' and Workers' Democratic Party (PW).

New parties, facing different problems, appeared in the 1930s. Early in the decade the Rural Reconstruction Association was established. Its leading figure was Liang Shuming, an intellectual concerned with village development.[5] In 1931 Carsun Chang (Zhang Zhunmai) and some associates formed the National Social Party. A few years later this group incorporated the Chinese Democratic Constitutionalist Party, which had first been set up in North America, and the name was changed to the Democratic Socialist Party. This conservative party consisted of a small group of intellectuals.[6]

In 1936, with the threat of Japanese imperialism of great concern to all Chinese, the National Salvation Association (NSA) was organized. A few months after its founding its leaders, the Seven Zhunzi,[7] were arrested by the Nationalist government for leftist inclinations, an act which inspired general public indignation. The NSA represented an effort to unite the nation against Japan, and it counted a number of Communist sympathizers among its leadership. It was the most popular of any of these small groups, which probably explains why it was the most persecuted by the government.[8]

Democratic League

These six disparate entities were not the only political parties that dotted the Chinese political scene between the two world wars, but they were the ones that banded together late in 1939 to form the United National Construction League (first known briefly as the Grand League of Democratic Political Groups). This league sat in the People's Political Council, along with Guomindang members and Communists. The United National Construction League underwent an organizational tightening, and in March 1941 it changed its name to the Federation of Democratic Parties. In October 1944 the first Congress of the Federation was held in Chongqing, and the name was changed once again, to the China Democratic League (DL). Relations between the Guomindang and the Communists had become increasingly tense since the founding of the original League in 1939, and the group (in its various permutations) came to look on itself as a mediating force. But in trying to assert itself, the DL antagonized Generalissimo Chiang Kai-shek (Jiang Jieshi), who, not entirely without justification, considered it to be dominated by Communists. Any League activist who operated openly was taking his life in his hands, especially after the war. (In 1946, after poet Wen Yiduo held a DL press conference in Kunming, he was gunned down.) Thus, operations often had to be conducted in secret and frequently from a base in Hong Kong, both before and after the Japanese occupation of that colony.[9] The atmosphere tended to have a radicalizing effect on the hitherto middle-of-the-road League members.

Although the Democratic League was a bourgeois rather than working-class[10] party, it was nonetheless a heterogeneous collection of autonomous groups--a factor that contributed to its impotence. It also contained individuals not associated with any constituent group.[11] Zhang Lan, who became its chairman in 1944, was such an individual. He was given the post, it appears, because he was generally respected by middle-of-the-road elements, at the same time being on friendly terms with the Communists and still in the favor of the Guomindang. Zhang, a candid, spirited man, had once briefly served as governor of Sichuan, and he had long presided over various Sichuan universities. Thus he was a significant leader who was not identified directly with the Guomindang government. He seems not to have been associated with any of the League's component groups, to whom he was willing to grant considerable independence. Zhang was thought of not as a spokesman for a particular political program, but rather as a proponent of national unity in the cause of winning the war against Japan and restoring peace in the country. Although he was nominally neutral vis-à-vis China's two great parties, the League moved steadily to the left during his chairmanship. Zhang would retain this post until his death in 1955.[12]

There are still questions regarding the League's activities during the war years. There are indications that it received financial assistance from the Communists.[13] And the Nationalists have claimed that in November 1944 the League promised the Communists that it would not cooperate with the Guomindang.[14] Choosing to believe the worst, the government acted in a highly repressive manner toward the League, thus alienating most of its members. The League came to see itself less a middle force than part of the opposition. This attitude was reinforced when, after the failure of the peace efforts of George Marshall, it tried unsuccessfully to continue the mediation attempt. Now, though policy differences with the Communists were recognized, it was deemed appropriate to minimize their significance. As one League spokesman rationalized:

> We quite agree with the Communist Party on China's need for democracy. Since the Communists modified their economic policies on the basis of definite recognition of private property, we agree with them on that subject. Their economic policies coincide with China's needs on the basis of existing conditions. . . . We are also of the conviction that the Communists will not revert to their more radical prewar policies when the war is over. As a matter of fact, although we have no assurances on that point, we believe that after the war the Communist Party may become even more democratic than it already is--providing the Communists are not again attacked by the Guomindang armies. . . . The Communists have no ambition of being the sole leaders of the nation. Of this we are convinced, and we know them very well. They are realists and know that the Chinese people will never really support and help a one-party dictatorship.[15]

It was easy to forget in 1945 that the Chinese Communist Party was largely a creature of the Soviet Union; it would be at least decades before the Stalinist influence could be shaken off, and even that would not herald the arrival of democracy.

During abortive postwar negotiations between the Communists and Nationalists to establish a coalition government, Mao Zedong insisted that the Democratic League be given six of the thirty-six seats in the Government Council. Because any major decision would require a two-thirds majority, he reasoned that a few non-Communists could be counted on to vote with the ten Communists to veto proposals.

In October 1945 the Democratic League held an extraordinary congress to elect a Central Executive Committee.[16] On this occasion, the incompatibility of the various factions became apparent. Hostility toward China's two militarized parties might suffice to bring the League's elements together, but it could not keep them together. Indeed, the conservative China Youth Party, which had dominated the secretariat, was now ousted from control, with the League falling under the grip of more leftist elements. Pro-Communist Zhou Xinmin became deputy director of the secretariat.[17] The coalition government idea was obviously getting nowhere, and by late 1946 the Guomindang had so alienated the League that most of its members refused to attend the National Assembly, which was held in November of that year. (Two of the League's member groups virtually severed their ties with the federation and took part in the assembly. These were the Youth Party and the Democratic Socialist Party, which still exist on Taiwan as satellites of the Guomindang.[18]) Thus the League lost the support of some conservative and moderate elements, making a rapprochement with the Nationalists less likely than ever.[19]

It was this more leftist League that participated in the Political Consultative Conference of January 1946, where, surprisingly, it had a plurality of delegates--nine out of thirty-eight. But the Nationalists arranged for the conservative Youth Party to have five seats, which put the Nationalists in control. The Communists only had seven seats, and at this point it was uncertain how much support they would have from the League. The Nationalists' maneuvering involving the Youth Party enraged most of the League, because the Youth Party technically was still one of its constituent groups. The League realized that the Nationalists were offering a disproportionate number of seats to the Youth Party (and also to the National Socialists) to induce them to leave the League and thus reduce it to a marginal entity. This disposed most League figures to lean still farther away from the government.

During 1946, the rightists who dominated the Nationalist Party seemed bent on harassing and repressing the League. They placed the group's offices under heavy watch. League luminary Luo Longji desperately urged the American diplomats in Nanjing to intervene on the organization's behalf. An embassy account of his concerns makes vivid reading:

> Dr. Luo's greatest concern stemmed from his conviction that the persons surrounding the League headquarters were members of the military secret police, and that in the event of arrest he anticipated secret action by military tribunals which in many past cases had resulted in the permanent disappearance of persons so arrested.
>
> In passing, it is interesting to note that Dr. Luo's fears in this connection are borne out to some extent by checking the registration number of the jeep which followed Dr. Luo to the Ambassador's residence and to the Chancery on October 23 [1947] and succeeding days. . . . The Embassy ascertained that the jeep which followed Dr. Luo . . . had been formerly registered with the office of General Dai Li [notorious head of the secret police]. . . . It was pointed out to Dr. Luo, however, that the League could not expect the United States Government actively to intervene on its behalf.

Thus, though some Americans sympathized with the League,[20] the U.S. government appeared to turn a deaf ear. Within days, the Nanking government's Ministry of the Interior declared the League pro-Communist and outlawed it, objections from Guomindang moderates notwithstanding. According to the official announcement,

> The Chinese people have known for a long time that the Democratic League has linked with the Communists and joined the rebellion. . . . In view of the seriousness of the Communist rebellion and the rampant activities carried on by the League, the Government can no longer tolerate an organization which opposes the National Constitution and

aims at the overthrow of the Government. For the preservation of peace and order in the rear, this Ministry has to take adequate steps to check the activities of the League. The Democratic League is hereby pronounced illegal.[21]

At this point, the League was given a choice. It could persist as an underground organization, or it could dissolve. In the latter eventuality, the government promised not to arrest former members except for overt acts. At first the leaders pointed out that they had no authority to dissolve the organization. They soon saw the handwriting on the wall, however, and announced the dissolution of the organization.

Suppressing the middle parties failed to "preserve the peace." Rather, it escalated defections to the Communists. The latter, by word and deed, were able to convince most moderates that their future as well as China's lay with them.

Many in the Democratic League denied the legitimacy of their party's dissolution. Some branches went underground or overseas. Chairman Zhang Lan defiantly went to Shanghai, where the Nationalist authorities placed him under house arrest (he was later spirited away by the Communists). But most of the leadership withdrew to Hong Kong. The following January (1948) a plenary session of the League's Central Executive Committee met in the British colony. Participants denounced the Guomindang and urged the Chinese people to rise up against Chiang Kai-shek and "liquidate" his party. A foreign policy statement was generally hostile to the United States. Finally, the League announced that it was willing to enter into a united front with the Communist Party and the Revolutionary Guomindang (RG).[22]

Revolutionary Guomindang

The Guomindang Revolutionary Committee, commonly known as the Revolutionary Guomindang, consisted of renegade elements from the Nationalist Party. The organization's roots can be traced to Nanchang in 1927, when leftists ousted from the Guomindang banded together in the face of Chiang Kai-shek's "white terror." After the failure of the Nanchang Uprising, the group became a quiet, largely underground intelligence-gathering organization.[23] After World War II, the RG merged with two organizations that had just been formed. These were the San-min-zhu-yi Comrades Association, led by Tan Pingshan and Chen Mingshu, and the San-min-zhu-yi Promotion Association under Cai Tingkai.[24] The Revolutionary Guomindang's mission was still largely sub rosa--gathering intelligence and persuading government functionaries and soldiers to defect. One former underground worker describes his assignment at that time as "sabotage, basically." He would seek out foreigners with whom Nationalist officials and generals associated and "find out what I could about strategies and plans through them. Inside Guomindang offices where I worked, I'd memorize secret documents, writing crucial figures on my hand."[25]

The Revolutionary Guomindang's overt existence was largely limited to Hong Kong, where it was reincarnated in the winter of 1947-48. At this time Li Jishen was made chairman, with Song Qingling as honorary chairwoman. Li, long an important political and military leader in South China, had been associated with the Guomindang since that party's early years. He became governor of Guangdong province in 1926; a year later, when the Communists attempted a coup in Canton, the provincial capital, Li brutally suppressed the uprising. Thousands died in the famous "Canton Commune," and it would be a long time before the Communists forgave Li for his actions.[26] He was also usually on bad terms with Chiang Kai-shek, who expelled him from the Guomindang in 1947. Li was accurately considered a Communist sympathizer.

Chiang had good reason to fear Li Jishen, who in 1948 came close to establishing a separatist government. Though he was now in self-imposed exile in Hong Kong, he had followers not only in South China but also in Beijing (then called Beiping). The thought was to carve China up into five regions, each under a semi-autonomous government. Li would be the titular leader, but only in actual control of Guangxi and Guangdong. By June 1948, Li was determined to establish a new government, but he hesitated because of what would happen to his followers in the capital. Nonetheless, he wanted to act before the U.S. presidential elections. American diplomats, who had been kept fully informed of his intentions, were asked to help extricate RG members in Nanjing. Whether the Americans were willing is not known, though they probably were not. They did not think highly of Li, but they did take him seriously. Plans were laid to have a diplomatic observer (Richard Service) in Guilin, the presumed location of the provisional government. In

July the Revolutionary Guomindang issued a manifesto in Hong Kong which stopped short of declaring a new government and was very pro-Communist in content. Only the local Communist paper published it.[27]

Events were moving too fast for Li Jishen. Soon his only option was to curry favor with the Communists, abandoning his intentions and facilitating their takeover. For their part, the Communists managed to have a selective memory as far as the Canton Massacre was concerned. They would shortly declare that Li's Revolutionary Guomindang was the "real" Guomindang, carrying the banner of Sun Yat-sen's Three Principles and Three Policies. Li became the leading personage of the reformed democratic parties in the People's Republic, but he was never given any meaningful political authority, and we are told that he always dreamed of his earlier days as a southern warlord.[28] His main role would be to personify Beijing's olive branch to potential Guomindang defectors on Taiwan.

National Construction Association

In December 1945 the National Construction Association (NCA) was formed in Chongqing. Unlike the other parties under review, the NCA became more significant after 1949. An outgrowth of Huang Yanpei's Vocational Educational Group (VEG), it was more politically oriented than its predecessor. It first operated, under Huang's leadership, within the Democratic League. Its purpose, like the League's, was to serve as a mediating force between the Communists and the Guomindang. It was driven underground in 1947 and eventually severed formal connections with the League. The NCA was designed to attract not only educational and cultural circles but, more important, members of the business community. The latter sector would eventually be its main focus.[29]

Jiusan

On September 3, 1945, members of a tiny leftist academic group called the Democratic Scientific Forum (or Society) met to celebrate the end of the war. They renamed their organization the Jiusan Study Society. The name, taken from the date of that meeting, literally means "nine three," which the society's detractors would jocularly twist into "nine members and three personnel."[30] Although an obvious exaggeration, this correctly suggests that the Jiusan has never been large.

Association for Promoting Democracy

In 1946, the Association for Promoting Democracy (APD) was organized. This was the same year as the founding of Cai Tingkai's similarly-named organization, the San-min-zhu-yi Promotion Association, and the early histories of the two groups are difficult to untangle. One source explains it this way:

> Because of the similarity in the names . . . the two were merged. But the Chinese Communists [later] wanted to maintain the names of many parties and have the opening of deliberations adorned by the presence of all these groups. In addition, Ma Xulun, the responsible leader of the APD, wanted a top position; he was unwilling to be placed under Cai. Nor would he accept the names "San-min-zhu-yi" or "Guomindang" (though Ma had held various posts in the Guomindang). For these reasons the APD was made completely independent [of the other minor parties].[31]

Peasants' and Workers' Democratic Party

Finally, there was the old Third Party, which was desperately seeking to effect a cease-fire between the Communists and Nationalists.[32] It soon changed its name to the Peasants' and Workers' Democratic Party, presumably a vain attempt to make the party more appealing to the masses.[33]

None of these parties appears to have had much mass appeal. This is not surprising, since China's mostly agrarian and often desperate people were largely concerned with immediate survival. The democrats may have been farsighted in the sense that they understood China's constitutional needs, but they had no way to popularize or carry out their programs.

As noted above, the Americans did not intervene on behalf of the middle parties. However, some influential Americans considered these groups China's only hope. General George Marshall, who spent more than a year in China trying to prevent civil war, saw Guomindang corruption as China's greatest problem, and one that no amount of foreign advice could rectify. Marshall believed that the key to cleaning up the government was an effective opposition party. Thus he urged that the middle parties should unify and form a counterweight against both the Guomindang and the Communists. But, as the State Department White Paper notes, "The minority parties . . . allowed themselves to be divided and were consequently unable to influence the situation or prevent the use of military force by the Government or the promotion of economic collapse by the Communists. In the midst of this deplorable situation stood the Chinese people alone bearing the full weight of the tragedy."[34] These parties did have their genuine fans, but there were also opportunists who would use the groups to their advantage.[35]

Acceptance of the Communist Victory

Whatever hope China had for democracy rested in these parties, which had reputations for integrity but few other assets. They were pathetic organizations, totally lacking in clout, with ill-defined purposes[36] and limited followings. Given the Communist victory in 1949, the logic of history would have suggested that they should enjoy merciful deaths. That they did not expire is less attributable to their stamina than to the cleverness of the Communists, who realized that these groups might still serve the ends of their Party. They had been a rallying point for Chinese opposed to dictatorship, but they were so weak they posed no threat to the Communists. Instead of destroying the parties outright, the Communists "surrounded" the liberal camp and brought it into their fold. This facilitated the consolidation of Communist control of the nation. In different ways, the efficacy of the technique was proven both before and after 1949. Initially, the result of the united front policy was to isolate the Guomindang, which was rapidly falling out of favor anyway due to its failure to cope with the political, economic, and military situations. Eventually, the policy would permit the Communist Party gracefully to control non-Communists, who wanted to (or had to) get along with the new regime. Thus, on December 25, 1948, Mao Zedong called for "the broadest united front" but insisted that "this united front . . . be under the firm leadership of the Chinese Communist Party."[37]

There was a somewhat contradictory emphasis here on "broadness" versus "firmness." During these months, the Communists more generally emphasized the need for unity among progressive forces. By convincing the middle parties that their own interests would be better served by siding with the Communist Party than by remaining independent or going over to the Guomindang, the Communists increased their own chances of a nationwide victory. In keeping with this policy, the 1948 May Day call urged that "all democratic parties and groups, peoples organizations, and social luminaries, speedily convene a political consultative conference, discuss and carry out the convoking of a people's representative assembly to establish a democratic coalition government."[38] Talks were held between the Communists and leaders of the minor parties both in Hong Kong and in Communist-held areas. In late 1948 agreement was reached at Harbin on the holding of a consultative conference. It would be called the Chinese People's Political Consultative Conference (CPPCC) and would comprise representatives of the twenty-odd organizations (among them the six democratic parties) who supported the May Day pronouncement.

In the summer of 1949 a CPPCC Preparatory Committee met to draw up rules for the conference, which was to convene in September. Also written was a "Common Program," which apparently went through seven drafts to meet the demands of the non-Communist groups. But all decisions were made under the guiding hand of the Communist Party, which made no important concessions. In the future, if any disagreements arose, the rules declared that dissidents would "have to faithfully abide by the . . . resolutions in accordance with the principle of the minority submitting to the majority."[39] By hand-picking all delegates to future CPPCCs, the Communists would assure themselves of virtually unanimous support.

The first CPPCC session was held in late September. Essentially, it was a demonstration of support for the Common Program. No dissent was heard. Like its successors, this session of the CPPCC served as a forum for Party-line speech making. After the conference ended, the headquarters of all the groups were moved to Beijing.

Impact on the Communist Rank and File

The united front strategy had been formulated by the top

Communist leadership, who did not intend to extinguish the middle parties. It is not likely that the Party rank and file wholeheartedly supported this policy. Indeed, at least one local Communist leader was sacked over this issue.[40] These partisans had waged a hard fight--many of them for decades--to establish a socialist state. Why now dilute that achievement unnecessarily? That the issue was a troublesome one within the Party is suggested by the number of times it received a firm answer. One came from the Party leadership as early as January 1949:

> The revolutionary camp . . . must embrace all persons willing to join the present revolution. The revolutionary cause of the Chinese people requires both a main force and allies. An army without allies cannot conquer its enemy. . . . There is no doubt that in China there are many loyal friends. . . . There is no doubt that not a single one of these should be forgotten or cold-shouldered. The raging tide of the Chinese revolution is forcing all strata of society to determine their stand. The relation of class forces in China is undergoing a new transformation. Great masses of people are now breaking away from the influence and control of the Guomindang and coming over to the revolutionary camp.[41]

Naturally, the purpose of such a statement was in part to reassure the minor parties. But it is the type of pronouncement that was to become familiar in another context--the recalcitrance of rank-and-file Communists in cooperating with non-Party people in ways desired by the leadership.

3
The DPGs and Their Role, 1950-1955

The first few years of the People's Republic of China constituted a period of considerable turmoil and brutality. Although this fact must be kept in mind when studying any aspect of the period, such happenings lie outside the scope of this book. The democratic parties and groups were redesigned for individuals for whom violent repression could presumably be avoided. The nation's new leaders believed that these groups, if led by the right people, would serve as a bridge (*qiaoliang zuoyong*), making it possible to reach key segments of society who otherwise would have closed minds.

The organizations were able to retain many--but not all--of their previous members, and the institutional names remained unchanged. Collectively, the groups continued to be known by their old generic name, democratic parties and groups (DPGs). The words "democratic" and "parties," however, took on altered meaning. Despite their appellation, these organizations are neither democratic nor parties in any commonly understood sense of these words.

Organizational Transformations

After the September 1949 CPPCC session, the Communist Party moved to tighten its grip over the DPGs. For most of the next fifteen years, their every activity--even topics of discussion at meetings--would be determined by the Party.

The job of reorganizing and supervising was the responsibility of a Party unit known as the United Front Department (UFD), which was also in charge of such strata as ethnic minorities and religious groups. For five years, the UFD had been headed by a Hunanese named Li Weihan. Li had studied in France during the early 1920s and had become an important member of the Communist movement after returning to China. A Long March participant, he might have ended up a member of the Party's inner core but for his habit of being on the "wrong" side of Party disputes. Nonetheless, UFD chairmanship was not an insignificant post. Li would hold it until 1965.[1] Under Li Weihan's direction, a general revamping took place within the old democratic parties. Members were organized locally into branches (primary units), or chapters.[2] These were usually organized around the workplace. Chapter leaders were told how they would have to conduct themselves; those who objected to the changes were either demoted to innocuous positions, given the status of ordinary members, or ousted. So many members were expelled that the DPGs appeared to lose totally their independent-liberal orientation.

Real anti-Communists, of course, might count themselves lucky if nothing worse happened to them than expulsion from their democratic party. But while these groups were not allowed to harbor irreconcilables, neither were they limited to individuals wholly won over to Communism. To be sure, counterrevolutionaries ("nonpeople") were denied any political status and could not belong to these groups. On the other hand, a person entirely reconciled to Communism belonged wherever he[3] could best serve the state--which might be in a minor party, in the Communist Party, or in both or neither of these. But the bulk of DPG members found themselves neither hundred-percenters nor counterrevolutionaries, and it was precisely the Communists' intention that the groups should be composed of people between those extremes.

The constitution of each DPG was revised during 1950, and several of the groups held congresses late in the year to ratify the changes. By now, the Communist Party was in full control. The case of the Revolutionary Guomindang, considered the senior DPG, was typical.[4] It was financed largely by public funds.[5] As a gesture toward tradition, the party was allowed to retain certain old organizational forms. It was to consist of four subgroups: the Min-ge (a Chinese abbreviation for RG), which claimed descent from the original organization; the San-min-zhu-yi Comrades Association; the San-min-zhu-yi Promotion Association; and the Fourth Subgroup, which comprised members not previously associated with any of the others. (There were some eyebrow-raising inclusions, such as a former imperial concubine.) Superimposed over all this, however, was the really meaningful structure. Where native "talent" was missing, the Communists supplied new blood. For example, Li Weihan's brother Li Junlong joined the party and immediately became a member of its Central Committee.[6] Several organs were established that radiated to the local chapters. The only organ calling for special mention is the Propaganda Bureau, through which the membership was to be informed of the demands of the state. One of the earliest requirements was that Marxist-Maoist thought be studied.[7]

By the end of 1950 the democratic parties had been sufficiently reordered to satisfy the Communist leadership. Now the minor parties could resume recruiting new members, which they had not been able to do for several years. But membership drives were to be governed by some peculiar rules. Under an arrangement that Premier Zhou Enlai appears to have been instrumental in fashioning, each group was allowed to enroll new members only from a special segment of society. Under this "sphere of influence" system, the domain of another party could not be encroached on--least of all that of the Communists. The DPGs were to restrict themselves to people deemed to have a potential to manifest bourgeois tendencies--that is, the more politically undesirable elements in this protosocialist state. Nor could just any bourgeois individual be recruited. Instead, it was required that each DPG restrict its activities to one segment of that class as determined by the Communists. The purpose of this system was to promote "efficiency."

Each democratic party did retain those members who had previously belonged to it and had not been purged. Thus, the Revolutionary Guomindang continued to count among its membership those who had once belonged to one of the three antecedent groups mentioned earlier, or to the Guomindang itself. But a new recruit had to have a particular occupation. In the case of the Revolutionary Guomindang, he usually had to have a position in the government.[8] The RG might also invite suitable individuals who took part in a campaign such as land reform or Resist-America Aid-Korea. Regardless of occupation, however, no one could join who was not a resident of a major city. Inhabitants of rural and ethnic minority areas did not normally join the Revolutionary Guomindang.

Four DPGs were designated for intellectuals. Such people, correctly assumed to be apprehensive about Communism, usually fell into the petty bourgeois category, which meant that they possessed bourgeois mentalities but were not necessarily property owners. The Democratic League was the most important party in this regard. It already had the largest following, most of whom could be classed as "intellectuals"--anyone with a college education or who possesses literate or technical skills. Now the League was to recruit exclusively from educational and cultural circles--particularly professors, school teachers, technicians, and university students. It did not actively recruit government workers; in particular, it avoided the army, police, intelligence agencies, and diplomatic corps.

The other three "intellectual" DPGs were smaller and were to remain that way. Their scope was not immediately defined with any precision. At first they were each to recruit from a broad range of intellectuals and professionals.[9] The Peasants' and Workers' Democratic Party was permitted to recruit members from a number of occupations, with only individuals in the armed forces specifically excluded. Gradually, however, the emphasis came to rest on the possessors of technical skills, especially doctors and others concerned with public health. As time progressed, the Association for Promoting Democracy would emphasize primary and secondary school teachers. The small Jiusan Society was to recruit among elite groups of scholars, scientists, and other highly skilled professionals.

Although businessmen were not excluded from the above DPGs, they were not actively recruited into them. "In accordance with our United Front missions," announced Zhang Bojun (of the DL and PW), "there is no further need to admit people from industrial and commercial circles."[10] The reason was that they had their own organization: the National Construction Association. According to chairman Huang Yanpei's account, on May 26, 1949 (two days after Shanghai was taken by the Communists), it was arranged by Zhou Enlai that henceforth the main responsibility of the NCA would be "to unite and educate patriotic industrialists and businessmen."[11] The Association would fill its ranks with selected individuals from among the "national bourgeoisie." According to the party's constitution, such people were "principally industrial and commercial capitalists and their agents." Included among those eligible were shopkeepers and active members of industrial and commercial organizations such as chambers of commerce. To facilitate the NCA's work, there would also be "an adequate number of the working personnel [as distinguished from capitalists] of government organs, public bodies and state-operated enterprises, the higher personnel of public-private joint enterprises, and private enterprises, as well as the intellectuals connected with industrial and commercial circles."[12] Unlike the Democratic League (for which recruitment was supposed to be limited to "key points"), the NCA was not required to restrict its recruiting to particular geographical areas. Indeed, a special provision was made in the constitution to set up sections where none had existed. It was expected that the organization would double or triple in size during 1951 alone.[13] Thus, under the guiding hand of the Communist Party, the National Construction Association was to grow from almost nothing into the second largest DPG. It would serve socialism in some very special ways, one of which was to "persuade" businessmen to turn their firms over to the state.[14]

Already it should be apparent that these groups were no longer political parties in any customary sense of the term. Their purpose was not to guide the affairs of state; this was the sole prerogative of the proletariat, for whom the Communist Party would speak. The DPGs certainly were not expected to be opposition parties; nor did they represent the genuine interests of their memberships, but rather the latter's "legitimate interests" (*hefa liyi*) as defined by the Communists. Later, the role of the DPGs would be even more frankly defined in *People's Daily*: "They should not reflect the bourgeois inclinations and demands of these people. That is to say, they should not represent those interests . . . which go against the laws of historical development."[15] The DPGs, then, would be parties in reverse. They were not supposed to have any impact on the nation's political life. But the decisions made in higher government and Communist Party circles would have a great effect on the lives of DPG members--perhaps all the more so because they belonged to these groups.

Although these internal political considerations probably explain the Communists' main motivations in keeping these small parties alive, the groups' existence had some additional propaganda value. The impression that the United Front would make not only on the domestic audience but also abroad was a real consideration. The Democratic League spokesman quoted earlier was naive but not altogether erroneous when he cited the Chinese people's distaste for one-party dictatorships. The Communists needed to make some gesture of conciliation, especially toward intellectuals and administrators. The services of such individuals would be crucial during the socialist construction, even if such people were not wholly sympathetic with this particular brand of socialism.

Furthermore, Chinese rulers traditionally, and the Communists in particular, have measured their acts in terms of the impact they might have on bordering regions not under the control of the central government. The most obvious instance pertains to Taiwan. For decades, the Communists have been talking about the "peaceful liberation" of the island, by which they have meant that if the Chiang family would capitulate, they and other officials would be given positions in the Beijing government. Revolutionary Guomindang spokesmen in particular have been active in reminding their ex-Guomindang brethren of this offer,[16] which a few have accepted.[17]

Also, it will be recalled that Mao and Liu Shaoqi were urging Communist parties in neighboring colonial and semicolonial countries to work to set up united front regimes. If the Communist Party were to go so far as to destroy the United Front even in name, the efforts of Communists elsewhere to imitate the first phase of the Mao-Liu strategy would be hindered by the demonstrated transiency of the united fronts being proposed.

Finally, the United Front has always been designed to appeal to overseas Chinese, many of whom have developed some appreciation of multiparty political systems.[18]

Activities in Government

The DPGs, as mentioned, are represented in the Chinese People's Political Consultative Conference. Not an effective decision-making body, the CPPCC serves as a forum for propagation of the Party line.

The various constitutions of the People's Republic of China have generally mentioned the CPPCC and the democratic parties in the preamble but overlooked them in the main body of the document. Neither are mentioned in the "Structure of the State" section of China's current (1982) constitution. When the first (1954) constitution was adopted, it was made clear that the CPPCC was being shorn of its original legislative role. With such functions largely taken over by the National People's Congress, the CPPCC would lack any real constitutional status and would simply act at the pleasure of the Communist Party. As Liu Shaoqi acknowledged in 1954:

> Some people have proposed that the status and tasks of the Chinese People's Political Consultative Conference should be specified in the Preamble. The Constitutional Drafting Committee believes there is no need to make such an addition to it. The CPPCC is the organizational form of our people's democratic United Front. It exercised its functions and powers in lieu of the National People's Congress and will, of course, no longer be required to exercise them in the future. It will, however, continue to play its part in the political life of our country as the organization of the United Front. Since it is the United Front organization, the parties, groups, and organizations in the United Front will, in consultation, themselves work out all the provisions concerning it.[19]

Thereafter, the CPPCC would not, either in theory or in fact, be part of the government. Rather, it is usually the place where approval of the line is expressed by all segments of the United Front. The Communist Party has the largest delegation in the CPPCC, though by no means a majority. Typically, there are forty avowed Communists and twenty-five representatives of each of the three most important DPGs. In the mid-1950s, these four parties, plus other groups represented in smaller numbers and individuals, made up a total of about 375 delegates.[20] Obviously, no one sits on the CPPCC, much less its central organs, without the prior approval of the Communist Party.

It is true that there have been non-Communists in the National People's Congress (NPC). Here, figures have sometimes been difficult to ascertain, since delegates do not sit as representatives of their organizations. But in the mid-1950s there may have been eighty-two members of the Democratic League in the NPC, and somewhat fewer from the other DPGs, as compared with 668 from the Communist Party.[21] A seat in the NPC, of course, still did not mean that a person had a real voice in the affairs of state.

This pattern would be repeated at all levels, with PPCCs and people's congresses at the provincial and local levels. Sometimes the Communist Party would even be in the minority in the congresses, but this was inconsequential inasmuch as the locus of political power lay in the Party and not in bodies such as these.[22]

Elections have usually followed the pattern typical for Communist countries (though there has been some liberalization in recent years). According to the system instituted in the 1950s, the Communist Party determines what names appear on the ballot--party affiliations are not shown--and then the entire population is urged to vote for the slate. The system has been frequently hailed by the leaders of the minor parties. Explained one Democratic League member: "We have consultations first and draw up a list of candidates. . . . As a member of the DL, I make suggestions. But I discuss these suggestions beforehand to avoid unseemly public disagreement. If there were no prior consultation, there would be a public controversy. This would waste time and the people's money, besides letting rich people exert their influence."[23]

A 1953 Revolutionary Guomindang announcement stated that the system "would greatly develop the people's activism and creativeness, and assure the thorough implementation of the general line of the state." Local RG units were ordered to "mobilize the whole body of members, especially those who are working in the government organs, to participate in the activities of voters' groups under the unified leadership of the local CP organs and the people's government."[24] Thus, the DPGs were encouraged to gain a sense of participation, but they were not in a position to affect the conduct of the Communist Party.

The Communists have always been well aware that there are many talented individuals among the "bourgeoisie" who have a contribution to make in the area of government administration. Though such people have often been given government positions, until the mid-1980s they were largely kept out of the Communist Party lest they contaminate its ranks. It has been considered more appropriate, therefore, for them to belong to the democratic parties. Non-Communists--mostly DPG members--have held various vice-chairmanships on the NPC Standing Committee[25] and during the 1950s headed a number of ministries. These ministries were usually concerned with sectors of the economy, such as light industry, food, communications, forestry, land reclamation, and water conservation. In addition, the redoubtable Shi Liang, a leading Democratic League figure who died in 1985, served as minister of justice until the abolition of that ministry in 1959. Various DPG members have held lesser posts in other ministries, such as Interior, Finance, Trade, Fuels, and Labor. Others, without portfolio, have been involved in helping to write the constitution and various statutes.[26]

The Communist Party leadership has often stressed the need for cooperation between Party and non-Party people in government. This was frequently the theme of speeches by Li Weihan. As early as April 1951 Li stressed the point, noting: "Non-Communists, in spite of their willingness to cooperate, have complained that they have been given the cold shoulder by Party members. It is up to the responsible Communist cadres of various organs to have this phenomenon rectified."[27] The strongest statement of this sort was made by Zhou Enlai on January 14, 1956, in a speech which that will be examined later in another context.

The frequency with which such statements were made during the 1950s is an indication that real cooperation was rarely achieved. But one need not rely on this evidence alone. During the "blooming and contending" that took place in 1957 it was revealed that non-Communists had frequently been hindered in their efforts to carry out their administrative tasks. Hundred Flowers as such will be discussed later, but it is appropriate to mention here a few of the difficulties about which non-Party government people complained. The big names in the DPGs, for example, were sometimes given more posts than one person could handle. In such cases, the actual work was done by others--usually Communist Party members. Indeed, nearly all DPG government people found that many important decisions were made by the Party without their being consulted. If decisions were supposed to be made by a body in which DPG people sat, the latter did so only as "guests." And although such administrators' responsibility was to act on the decisions thus made, often they lacked the authority to do so effectively. Their subordinates, if Party people, looked primarily to their CP superiors for orders. From the latter they obtained the clearest idea of what was expected. Channels of communications between the Party and non-Party administrators were often bypassed. Information on directives was frequently classified and available only to Party members. If the inevitable ignorance of non-Party administrators sometimes caused them to make mistakes, it was hardly their fault. Nevertheless, they were often blamed for failures.

Thus the United Front in government did not always live up to the expectations of the Communist Party leadership. It certainly fell short of the hopes of the DPGs.

Campaigns and Reforms

During the early years of the People's Republic the Communists launched various drives to change Chinese society. Some of the methods resorted to were brutal, and many people, including members of DPGs, lost their lives.[28] But more common were the mass campaigns and class reformations. The purposes of these were similar to those of the more blunt measures, although the targets and techniques differed. Mass campaigns were aimed at various segments of Chinese society, the DPGs being only a few of the groups participating.[29] Each reform targeted a specific segment of society. The reforms of interest here dealt with the bourgeoisie.

The first campaign in which the DPGs were obliged to participate was Resist-America Aid Korea. After the government's "June First" (1951) call for patriotism, arms donations, and relief aid, the DPGs took up the cry. The Democratic League adopted the campaign as a "central task" and called on all members to "insure the fulfillment of donations."[30] Similar efforts were made by other parties.[31] The uniformity of DPG announcements in content and timing is striking. Obviously, such actions were in response to an order of the United Front Department. Sometimes the statements were made jointly by several parties, as was the one in March 1952 condemning American conduct in the war.[32] When the various DPGs did issue separate statements, they tended to be similar if not identical.

The drive against counterrevolutionaries and the Three-Anti Five-Anti campaign[33] were of more direct concern to the DPG community and probably produced more tensions. It is clear from statements made in 1957 that much bitterness was engendered.[34] In particular, members of the National Construction Association[35] must have been disillusioned. These businessmen had been led to believe that they would be largely exempt from "socialism," but in the Five-Anti campaign they learned that they were subject to the heavy hand of the state.

Beginning in the spring of 1951, democratic party members

were told that, in order for their members to make a greater contribution to the socialist construction, people would have to shed their old ways of thinking. These organizations were to help this process along. For the next few years, "unity and reform" of intellectuals and business people was the dominate theme of DPG activities. Later in 1951 the Central Standing Committee of the Democratic League outlined the procedures to be followed. Members were required to participate in a far-reaching ideological reform campaign. "Criticism and self-criticism should be actively developed, thereby correcting various erroneous and even reactionary thinking among our ranks." League units in a number of universities were proudly singled out as instances where the decision had already been made "to make teachers' study campaigns the most important political task."[36] Similar statements appeared during the following years, and as late as January 1956 there were reports of the assistance the League was giving its members in solving their "difficulties" in work and study.[37]

Parallel developments were taking place within the Revolutionary Guomindang. According to a New China News Agency dispatch, "Some undesirable elements have been purged and ideological re-education has been conducted within the organization." Such "successes," the Revolutionary Guomindang acknowledged, "are mainly due to the correct leadership of the CP and the implementation of the correct directives from the RG Central Committee by the rank-and-file members."[38] But the "successes" tended to be overstated and often were counterproductive, as one RG member made clear in 1957:

Some say: "When a yamen servant is at fault he is beaten, and when his superior is at fault the servant still gets beaten." So it is with non-Party cadres, who thus become very frustrated. Applying big caps [labels] is a frequent occurrence; caps such as "impure motives," "capitalist class working style," "jeopardizing revolutionary enterprises," "undermining the prestige of the Party," etc., have been put on many non-Party cadres. As a result, people just try to stay out of trouble, and go along with what they know to be wrong.[39]

As with other professionals, there was much resentment at

having to take orders from unqualified supervisors.[40]

The National Construction Association was working diligently to rescue the national bourgeoisie from its backward tendencies. It launched its "ideological remolding movement" in January 1951, which seems to have been a few months earlier than the other DPGs.[41] The standard techniques of "self-examination" and "criticism and self-criticism" were to be employed.[42]

The efforts of the personnel in a Shanghai pen manufacturing concern are illuminating. In January 1956, nineteen individuals submitted personal self-reform plans to the local Construction chapter head. According to the New China News Agency:

> These plans have undergone six discussions and revisions . . . . The individuals have included their personal work, learning, ideological style, and family life. Some planned to read within a year textbooks on political economics and histories of the Soviet Communist Party. Some planned to arrive at the office promptly every morning according to the starting hour of the enterprise. They are assured of mutual supervision in the implementation of every item in their plans.[43]

The NCA was not always successful in bringing their charges around. During Hundred Flowers, we learn that many businessmen believed that they had been cheated out of their enterprises and wanted to be compensated. They also resented the fact that their managerial skills now went largely untapped.[44]

Still, efforts to remold such people were more successful than similar efforts among the intellectuals. It was the latter who were to manifest the most pronounced "bourgeois" tendencies when the pressure came off in 1957. Although there is no way of making a precise evaluation, the statements of top Communist Party leaders can be evaluated to see how far they thought the bourgeoisie had progressed. Several estimates were made during 1956, and it was largely on the basis of these that the Party encouraged the "blooming and contending" that took place in 1957.

In January 1956, Premier Zhou Enlai made two important speeches on the question of intellectuals. In one, delivered on January 14, Zhou told a large group of important Party personages that the intellectuals needed better leadership and more cooperation. Most intellectuals were now in government service and were "already part of the working class." Still, Zhou found it necessary to acknowledge that not more than 45 percent of them were definitely progressive. Ten percent were "backward," and there were even a few counterrevolutionaries to be counted among these. The remainder of the intellectuals were "middle of the road." Able to fulfill their tasks in the service of socialism, they were still insufficiently progressive, and further reeducation was called for. Some aspects of Zhou's speech must have been quite well received by intellectuals. There was an indirect hint that Western learning might once more be respectable.[45] In letters, and also in science, Zhou called for less reliance on the Soviet Union. An intellectual should be respected, the premier implied, even if his or her political outlook was somewhat backward. If "wrong-thinking" individuals "do not turn against the people in speech and action, and, even more, if they were prepared to devote their knowledge and energies to serving the people, we must be able to wait for the gradual awakening of their consciousness and help them patiently--while at the same time criticizing their wrong ideology."[46]

Two weeks later Zhou addressed a unit of the CPPCC. Not only did he now have some kind words for the intellectuals, but he also made a favorable mention of the DPGs. "In the main, the forces of the country's intelligentsia have been mobilized under the leadership of the CP and the government. And in this respect, the democratic parties and groups have done a good deal."[47] This, of course, was not the kind of praise that the intellectuals would have preferred, but it was better than none. In the two speeches Zhou made disparate assertions (with many sops to the leftists), and it is difficult to know exactly how to interpret them. But the novelty--and significance--lies in his conciliatory attitude toward the bourgeois intellectuals.

Later in the year Li Weihan also sounded optimistic in discussing the bourgeoisie and the progress that class had made. Among the intellectuals he noted fundamental changes, and he reiterated what Zhou Enlai had said about so many of them now being part of the working class. Capitalists had also taken big strides, and their relations with the workers were becoming imbued with the socialist spirit of "working together."[48] However, this was not seen as a moment to relax. Many intellectuals were still unable to escape their bourgeois backgrounds. Among the national bourgeoisie, remnants of capitalism still existed.

Liu Shaoqi likewise felt that the bourgeois intellectuals had a long way to go before they would be good citizens in the new order; nonetheless, he held a constructive view of the democratic parties, which he thought might even do a better job of mobilizing these people than the Communist Party. He implied, in remarks aimed at the DPGs in 1956, that the emphasis should be less on reforming the members and more on mobilizing the intellectuals. "Without you, our strength would be less, and many positive factors would not be mobilized. Of course, if you were alone and without us, you would not be able to mobilize much more. We can mobilize some, and you can mobilize others."[49]

So while some leaders at times played down the urgency of transforming the bourgeoisie, the need was never lost sight of, and it was frequently stressed. Said *Guangming Daily* on March 3, 1956: "The democratic parties must supervise their members to rise and lead the masses in participation of the criticism of the bourgeois academic ideology. . . . Doing a good job in this will be of great value for the self-reform of intellectuals, particularly higher intellectuals.[50]

The picture that emerges of the democratic party is thus one of a group of men (very few women), some of whom are wholly devoted to socialism, and some who the Communists feel will make especially fast "progress." The DPGs did not contain the whole subclass, or even a representative cross-section of it. Rather, to use the Chinese expression commonly used in the 1950s, they contained "backbone elements" (*gugan fenzi*), elements which could be counted on to pull the whole subclass to the left.

Stripped of their independence and any opportunity for spontaneity, the DPGs had become insipid organizations. Although they were supposed to have a "consultative" relationship with the Communist Party, in 1956 one Democratic League leader could not give a single instance in which the DPGs had influenced the Communists to revise a projected policy.[52] All that members could do was "respond to summonses." The interest level was so low that members often failed to appear for meetings. There was a sense of futility and boredom.[52] But things were about to change.

4
The Year of Transitions, 1956

The year 1956 was a critical one in the history of the DPGs. At this time the Communists attempted to define the role of these groups and offer a rationale for their existence. It was the first serious effort along these lines since 1945, and it was hoped that the new formulation would last for many years to come. Although it would be severely shaken in the following year and again during the Cultural Revolution, it still stands.

A reassessment of the role of the DPGs was necessary in 1956 because of the perceived transition taking place in the nation that year. This was the moment of the "basic achievement of socialism." Although China was not yet industrialized and classes still existed, the year was seen as a turning point. With the state virtually in control of the means of production, it could now be said: "The system of exploitation has been basically abolished, and the socialist system has been basically established."[1] Somewhat incongruously, however, there was no suggestion that the bourgeoisie should be declared extinct, nor that its "parties" should be abolished.

"Long-Term Coexistence and Mutual Supervision"

The new basis of the democratic parties was the assumption that bourgeois *mentalities* would outlive the bourgeoisie as a social class. The new role that these groups were to play was summed up in an often-repeated phrase that first became current in the summer of 1956: "Long-term coexistence and mutual supervision" (*changqi gongcun, huxiang jiandu*). As one writer put it, this concept "confirmed the important role played by the democratic parties . . . in uniting and mobilizing all active forces of the nation in our fight against internal and external enemies."[2] It did more than this, however. It assured the DPGs that their lives would not be coterminal with the existence of the bourgeois class. And it gave them a glimmer of hope that they might have some genuine impact on the political life of China.

Inasmuch as the democratic parties were organs of various segments of the bourgeoisie, and members of that class were soon expected to complete their remolding and join the ranks of the proletariat, what explanation was to be given for the continued existence of the DPGs? During the first few months of "long-term coexistence" there was no answer--which may have caused some uneasiness among the circles who were told to discuss the subject at their study meetings. But at the Eighth Communist Party Congress, Liu Shaoqi provided an explanation. Said Liu: "Since survivals of bourgeois ideology will long linger in the thought of this [former bourgeois] segment of the workers, there will long be a need for the DPGs to keep in touch with them, represent them, and help them remold themselves."[3]

Liu spoke to the congress on September 15. Ten days later Li Weihan gave his report, "The United Front Work and the Party," in which he elaborated on the Party position. Ideological changes, Li noted, often lag behind changes in social status. So it would be with the bourgeoisie. After the latter had ceased to exist as a class, the bourgeois mentality would persist among individuals. Work of "uniting, education, and transforming" would still be called for.[4]

The central organs of the DPGs ordered their rank and file to study the documents that came out of the congress. The Central Committee of the Revolutionary Guomindang, reportedly "jubilant in the extreme," ordered thorough discussions on all levels.[5] Other DPGs, said to be equally enthusiastic, followed suit.[6] But how long, DPG members must have wondered, is "long-term" coexistence? Li Weihan simply said "for a very long time," and others gave similar vague answers. According to Mao Zedong, however, the Chinese Communist Party, like the state, was eventually supposed to wither away.[7] There were certain hints, though not from top Party theoreticians, that the "witherings" of the CP and the DPGs would coincide. In July, *People's Daily* stated that "So long as the Communist Party exists, the other democratic parties will continue to exist."[8] And later in the year, a writer in the official Party journal *Study* referred to the Communist Party and DPGs "living and dying together."[9]

Though there was nothing new in the idea that Communists should supervise others, the idea of "mutuality" was a milestone. This was a special application of the principle that the Party needs the support and guidance of the "broad masses," and it had been foreshadowed by Mao's earlier assertions. China's bourgeoisie was now seen as being a legitimate part of the masses, and in some ways the most sophisticated part. The democratic parties, Li Weihan acknowledged before the Eighth Party Congress, represent people "possessed of certain experience in the political and economic fields and professional skills as well. Often they put forward to us well-directed criticisms and valid ideas. Although some of the criticisms may not be well-directed, and some ideas may not be so valid, still they will be of use to us in analyzing and dealing with various problems and help us keep a cool head. We must create all the necessary conditions to facilitate their supervision over us." Thus it would be desirable for bourgeois personages to play a greater role in government. Li promised them real work to do, and sufficient authority to carry out their tasks. And CP personnel in all organs were reminded that they should have "full consultation with the personnel outside the Party through appropriate channels."[10]

Difficulties in carrying out "mutual supervision" stemmed not only from the indifference of Party members, but also from the fact that there was no adequate machinery through which it might be realized. Conceivably, the CPPCC might serve this purpose. "The Central Committee of the Party and Chairman Mao attach the greatest [importance] to the CPPCC," said Li. "We should strive to further promote and enliven its activities."[11] In reality, the delegations of the DPGs in that body had rarely been encouraged to be anything but totally subservient to the Communists there.[12] Thus, non-Party people could be effective only in a few isolated administrative posts.

At the time, it must have been difficult for the DPG community to know exactly what the Party had in mind. With hindsight, however, it is possible to reconstruct the Communists' attitude in promulgating the doctrine of long-term coexistence and mutual supervision. Basic to their thinking was the confident belief that the bourgeoisie was making progress, that it was becoming reconciled to socialism, and that individuals not so reconciled could be held in check. Many elements within the bourgeoisie, it was felt, were in a position to make a valuable contribution to China's modernization. The CP would accept assistance and constructive criticism from such elements. But long-term coexistence and mutual supervision did not invite opposition. Instead, it was predicated on the belief that opposition would not be forthcoming.

Organizational Changes and Membership Expansions

Around 1949 the democratic parties and groups were of small, often indeterminate size and were shrinking. Altogether, there could not have been more than a few tens of thousand members.[13] In the mid-1950s they were permitted to expand. Among the largest increases was that enjoyed by the NCA. Between 1945 and 1955 the Association expanded fortyfold, albeit from tiny beginnings. The Revolutionary Guomindang, it is said, expanded two and one-half times by 1955. The Democratic League, the largest democratic party, had a less impressive percentage increase. Nonetheless, it managed to expand "276 percent" between 1950 and 1953 (at a time when many liberals were being expelled), and by July 1956 it had expanded three and one-half times. The other parties also grew during this period.[14]

Thus the DPGs expanded gradually during the 1951-56 period to a total membership nearing 100,000--doubtless more than the parties had had in the late 1940s. Still, over this number of years the increases were not on so great a scale as to revolutionize the nature of the organizations, which were becoming attuned to the political status quo. Admissions policies were always laid down by the Communist Party, to assure that the groups' complexions would remain consistent. Between the summers of 1956 and 1957, however, the picture changed. At this time the Communist Party slackened its control over the DPGs, and in some cases people who were later called "rightists" gained the ascendency.

Before discussing the changes that took place in the memberships, it is necessary to examine certain organizational changes. These took various forms. In the Democratic League, Zhang Bojun and Luo Longji organized four teams, which, the Communists later determined, were designed to "harm" the League.[15] And in the Jiusan, alterations made by Xu Deheng are said to have "seriously undermined the system of democratic centralism within the Society, causing alienation from reality and the masses."[16] Similar changes appear also to have taken place at the local level, where putative "rightists" sought to control party organizations[17] and to set up supposedly anti-Communist subgroups.[18]

Some DPG leaders also worked for qualitative changes in the memberships of their organizations. For example, Luo Longji later confessed that the Democratic League had been placing too much emphasis "on the absorption of middle-of-the-road and backward elements as members, causing a sharp rise in the numbers of such elements in the League."[19] Or, as Li Weihan would claim, in the NCA "some people twisted the principle of recruiting members mainly from the middle and upper strata, and wanted the big capitalists to manipulate and control the Association."[20]

There were serious disagreements in the DPGs among those who were trying to work out recruiting policies. Aside from those who wanted to carry out the wishes of the CP--and there were always many such people around--some sought to do away with the occupational sphere-of-influence system, but others wanted to continue it in a modified form. Zhang Bojun was among the latter. He sought a much larger League, but he wanted it to remain exclusively for intellectuals. The League, in fact, and not the CP, was to be their party.[21]

In this, Long Yun concurred.[22] And Luo Longji apparently wanted to see a merger of the four parties that had been primarily oriented toward the intellectuals--the League, the Jiusan, the Peasants' and Workers' Party, and the Association for Promoting Democracy--a plan the Communists themselves had once considered. Obviously, this would have consolidated the strength of the non-Communist forces.

The move toward consolidation was unsuccessful, but efforts to expand the DPGs were fruitul. Zhang Bojun even looked to a total DPG membership in the millions.[23] As he later confessed,

> I expected each democratic party to recruit several hundred thousand members and several democratic parties to recruit one or two million members. I also advocated expanding our organizations to the county level, and suggested that DPGs hold consultations on the recruitment of members in the counties. I said once, "after the basic completion of the

socialist revolution, the character and task of the democratic parties should be reevaluated and should be placed on a more advanced plain."[24]

A number of DPG leaders acted on such a line of thinking. Huang Jixiang sent a letter to the PW's Tianjin section (and no doubt others) advising them to "expand rapidly," making full use of the "current favorable conditions."[25] And the Jiusan's Central Committee informed local organs that "in membership expansion work it will not be necessary to refer everything to the [local CP] committees for decision"[26]--an indication of the way the Communist Party had heretofore determined who was to be a member of a DPG.

Although only limited (and somewhat conflicting) data are available, it appears that in some parties radical membership increases took place in 1956 and the first half of 1957. The best information is for the Jiusan, which increased 235 percent during this period and encroached on populations that had been reserved for the Revolutionary Guomindang and the Democratic League. The Society even enrolled some "workers," who were only supposed to be interested in the Communist Party.[27] By mid-1957, more than ninety local Jiusan organizations existed, whereas a year earlier there had been only eighteen-- and even these had grown tremendously. *Guangming Daily* reported that in Hangzhou, Jiusan members increased from 5 to 183, and in Jinan, from 4 to 180; in Tianjin, membership increased ninefold, and in Tangshan, twentyfold.[28] (These figures are of interest not only for the growth indicated, but also because they suggest that by 1956 in some areas the Jiusan had almost died out.) Local DPG chapters were often established even where there was no provincial organization.[29]

The figures are less complete for the Peasants' and Workers' Party, but the group seems to have paralleled the Jiusan pattern. It began recruiting people outside of health circles, which it had not been supposed to do, and it even branched out from the cities into rural areas. Zhang Bojun, its chairman, is said to have wanted to increase the membership to ten thousand.[30] In Tianjin it is said to have been easier to join the PW than to purchase a movie ticket.[31] And in Wuhan the party enrolled three hundred new members, which was half again the "original membership."[31]

Both the Jiusan and the PW had always been small in comparison with the Democratic League, Revolutionary Guomindang, and National Construction Association. Although expansions took place within these larger parties, few figures are available, and they are mostly of a local nature. In Shanghai, under the influence of policies attributed to Zhang Bojun and Luo Longji, the Democratic League grew from 967 (early 1956) to 3,384 (August 1957).[32] Nationwide, however, League expansion appears to have been largely a matter of intentions rather than achievement.[33] The picture of the two other larger DPGs is obscure, but there is fragmentary evidence of substantial increases in membership. In Wuhan, the Revolutionary Guomindang increased 50 percent over the early 1956 figures. In the same city the National Construction Association increased from under 400 to over 1,200.[39]

In short, the expansion of the smaller democratic parties was quantitatively greater than, and qualitatively different from, earlier periods. This does not appear to have been true of the big three, whose relative increases in size were more or less in line with those of the early 1950s.

Certainly these groups could not have grown like this (and in some cases even change the nature of their enrollments) without the knowledge of the Communist Party. Did the Party underestimate what was happening? Did it know what was happening and simply tolerate developments? Or might it even have approved and encouraged these? While no definite answer can be given, the latter alternative should not be dismissed out of hand. It is hardly a coincidence that it all happened after the doctrines of Hundred Flowers and long-term coexistence were declared. Perhaps some Communists realized that the United Front was becoming an empty fiction, and that the democratic parties would have to be enlarged if they were to be useful. If it was foreseen that the DPGs might develop opposition tendencies, perhaps some in the Communist Party were interested in determining what bourgeois elements would be attracted to them under such circumstances. If the Party leadership really expected the expanded DPGs to continue on their "progressive" paths, they apparently failed to appreciate the fact that liberals were still basically unsympathetic with communism, and that any expansion of the DPGs without a commensurate reinforcement of outside controls was bound to have untoward consequences.

5
The DPGs and "Hundred Flowers," 1957

The "blooming of the hundred flowers"--the brief letup of the pressure to which China's intellectuals had been subjected, had a major impact on the democratic parties. Many of the people who took the opportunity to air their criticisms of the Communists were members of these groups. In their Hundred Flowers activities they operated in this capacity, seeking publicity through DPG forums and publications. And some even tried to use the DPGs to gain political advantage, which implied altering the very nature of these organizations and turning them into real political parties once again.

"Contending"—The Underlying Philosophy

It is difficult to know exactly when Communist leaders began thinking in terms of permitting freer speech by intellectuals, but in the first half of 1956 a shift in attitude could be detected. Until this time the emphasis had been increasingly on repression, transformation, and mobilization. Such were the themes at the DPG higher-organ conferences in December 1955. Nothing seemed less in the air than "blooming" and "contending."

Zhou Enlai's speeches the next month, however, suggested a more conciliatory attitude. For the time being, Zhou was willing to accept the intellectuals as they were, instead of pressing for an immediate transition into good socialist men and women.[1] The next few months undoubtedly saw considerable discussion, and perhaps controversy, at the higher Party levels. In a fascinating secret talk, Mao Zedong stressed the merits of a multiparty system. A two-thirds cut in Party and government personnel which he was advocating did not appear to be applicable to the DPGs, toward which he seemed to have an odd but genuine respect: "We have a bunch of democratic parties, quite a few of which have their own opinions about us. Toward these people, our policy is unity plus struggle, because we want to mobilize them to serve socialism. Opposition [parties] do not formally exist in China, because all of the DPGs accept the leadership of the CP. In actual fact, some members of the DPGs form an opposition." These people, Mao noted, had equivocated during the civil war, the Korean war, land reform, and the campaign against counterrevolutionaries: "They oppose and yet don't oppose. Their patriotism often transforms them from opponents into supporters. Now, the relationship between the CP and the democratic parties needs to be improved. We must allow the DPGs to express their own views. As long as they are reasonable, we can accept their views regardless of whose views they are. This would be a rational [approach] for the Party, the state, and people, and for socialism."[2]

On May 2, Mao Zedong made a less secret announcement on the general subject of intellectuals. Speaking before the Supreme State Conference (which included some non-Party people), Mao set forth his plan for a liberalization, borrowing the term "hundred flowers" from ancient Chinese history.[3] The talk was not published, but its main features can be inferred from a speech made later in the month by Lu Dingyi, director of the Party's Propaganda Department.[4]

Lu acknowledged that his remarks were prompted by Mao's May 2 speech. The general theme was that China would benefit from freer political discussion. But his exhortations fell somewhat short of carte blanche. "The CP Central Committee . . . has pointed out that when criticizing the mistaken thoughts of the bourgeoisie and discussing academic problems, the Party's United Front and the policy to unite and reform the intellectuals must both be firmly adhered to." With such a reminder it is small wonder that China's intellectuals and businessmen hesitated for almost a full year before taking Hundred Flowers seriously.[5]

The concept, however, was not allowed to die during the intervening months. It was mentioned at various assemblies of the DPGs in the summer of 1956, at the same time that long-term coexistence was so much discussed. It was also reiterated at the Eighth Communist Party Congress in September.[6] And it was a key theme in Mao Zedong's important statement in February 1957. "On the Correct Handling of Contradictions Among the

People," like Mao's speech of the previous May, was delivered before a closed session of the Supreme State Conference, was heard by some non-Party people, but was not immediately made public.

Mao had first outlined his thoughts on "contradictions" (*maodun*) many years earlier.[7] The idea is closely related to the Hegelian dialectic, though it is different from the Soviet formulation of that concept. Mao had acknowledged that the "dialectical world outlook" could be traced back to ancient times in both East and West--presumably having in mind *yin* and *yang* for the case of China. "Ancient dialectics, however, had a somewhat spontaneous and naive character; in the social and historical conditions then prevailing, it was not yet able to form a theoretical system, hence it could not fully explain the world and was supplanted by metaphysics."[8] Hegel is given credit for making "important contributions" to dialectics, but the real strides are seen as having been made by Marx and Engels with their concept of dialectical materialism. For Mao, there was a distinction to be made between qualitatively different types of contradictions. Normally, contradictions between the bourgeoisie and the proletariat could only be resolved by revolutionary means. On the other hand, contradictions within the Communist Party are solved by criticism and self-criticism. These are the extremes; in dealing with intermediate situations one should not be dogmatic. "Processes change, old processes and old contradictions disappear, new processes and new contradictions emerge, and the methods of resolving contradictions differ accordingly."[9] Furthermore, a distinction was to be made between contradiction and "antagonism" (*duikang*). The latter, but not the former, would disappear under socialism.

This point was expanded in Mao's 1957 speech. Now, two types of contradictions were seen: antagonistic and unantagonistic. Antagonistic contradictions exist within a capitalist society and between capitalist and socialist societies. But "contradictions in a socialist society are . . . not antagonistic and can be resolved one after another by the socialist system itself."[10] And now that socialist relations of production had been established, it was desirable to modify the tactics used in the domestic class struggle. Nonrevolutionary forces could be dealt with in the open; persuasion was to replace the use of force.

Mao exhorted the intellectuals to be more outspoken. He felt able to do so because "as a scientific truth, Marxism fears no criticism."[11] He realized that freer speech would facilitate "the flourishing of the arts and the progress of science," but most of his remarks also had political implications. What Mao really expected Hundred Flowers to accomplish was to increase the strength and vitality of the Communist Party. He likened the struggle against "wrong ideas" to a person's being vaccinated. One is healthier after having fought off the few bad germs that have been placed in one's system. Mao did not mean, of course, that counterrevolutionaries or "wreckers of the socialist cause" should be heard from. But among the "people" (which in such a context excludes counterrevolutionaries, though not the whole bourgeoisie), it was not desirable to ban the expression of "incorrect" ideas. "It is not only futile but very harmful to use crude and summary methods to deal with ideological questions among the people. . . . You may ban the expression of wrong ideas, but the ideas will still be there."[12] This may sound like a restatement of a cardinal principal of Western liberalism. Indeed, Mao professed to be offended by the notion of suppressing ideas, which he saw as a poor way to eliminate contradictions. But ideological pluralism was encouraged not so much to insure that the truth would be laid bare as to make the errant individual look foolish in his own eyes.[13]

The motivations underlying Hundred Flowers, therefore, were twofold. One aspect harks back to "mutual supervision," a concept that is given due emphasis in Mao Zedong's pronouncements.[14] In May 1957, furthermore, the idea that Communists might have something to learn from non-Communists was given some meaning in practice. Why, at this time, there was such an emphasis on "mutual supervision," almost to the exclusion of the other facet of Mao's thinking--the need to draw more intellectuals to the left--is still difficult to determine. A superficial analysis might lead to the conclusion that Hundred Flowers was simply a trick played on intellectuals.[15] But this assumes that the events of the spring of 1957 were anticipated by the Party. It also overlooks the Communists' genuine idealism, and optimism that the intellectuals were making progress. However unrealistic these attitudes may have been, they were clearly reflected in Mao's speech. It is unlikely that he would have made such a public error in judgment if he did not believe what he was saying. So while Hundred Flowers ultimately put the Communist Party in a new position of renewed dominance over China's intellectuals, this fact of history should not obscure the probability that the Party leadership *expected* some positive benefits other than this to accrue to their larger cause from mutual supervision.

Blooming and Contending

Although there were certain rumblings of activity and further definitions of policy during the next two months, the only development worth noting is the launching of the rectification campaign within the Communist Party. This was announced in a *People's Daily* article by Lu Dingyi shortly after Mao's February speech. It had been fifteen long years since the original rectification (1942), in spite of the fact that the Party membership had multiplied more than twelvefold. It was believed that some members could use further ideological training and general disciplining. In particular, many cadres were rustic products of guerrilla warfare. Now that China was trying to modernize, they had much to learn from China's more refined, better-educated people.

Thus, the peculiar feature of the 1957 rectification was that it was not entirely self-contained within the Party, but instead help was invited from the outside. Toward this end, on May 8 Li Weihan convened the first of two forums in which all the DPGs participated.[16] It was at these May Forums that much of the famous "blooming and contending" took place. Not only did Li announce at the first session that the purpose of the forums would be to aid the Communist Party in its rectification, but it was emphasized that participants were to criticize the Party freely.[17]

Such criticisms were soon forthcoming. They were made at the May Forums, in the DPG press, and at many democratic party meetings. While care was usually taken to direct remarks at the Party *membership* (as distinct from the institution), this did not seem to blunt the attacks. Nor was this the effect of efforts to express grudges in terms of acceptable clichés, such as "walls" and "ditches" (within the United Front) and the "three evils" (bureaucratism, sectarianism, subjectivism). Such gestures often became crude burlesques, as when Zhang Naiqi announced: "I consider bureaucratism . . . a more dangerous enemy than capitalism."[18]

Constant themes were the shortcomings in CP-DPG relations, and in the relations between Party and non-Party people in government. At the meeting of the RG Central Committee, for example, Qian Zhangzhao complained: "Individual CP members are arrogant. Sometimes when they are contacted on official business they neither reply to letters nor answer telephone calls, and after repeated reminders they give you a cool answer to your question."[19] On another occasion, Hu Zi'ang (NCA) cited unequal treatment in job assignments and remuneration.[20] And a member of the Peasants' and Workers' Party challenged the claim that Communist Party members were the first to bear hardship and the last to enjoy comfort.[21]

Frequently heard was the cry that the role of the Party was too great and that of the state too small. According to the RG's Shao Lizi, the government had no power, for it was all in the hands of the Party. The latter was seen by a League member as having failed to distinguish between itself and the government. It ought to abide by the "existing laws, regulations and systems."[22] But usually it was the *lack* of any legal system that drew complaints. It was pointed out that there was no code; instead, the public had to put up with the high-handed ways of "political aristocrats"[23] in the Party. Unless a legal system was established, argued Li Renren (RG), the rectification could not achieve any lasting success.[24] Democratic League member Yang Qing observed that party members at all levels "habitually flout the Constitution," and he had a suggestion: "I propose that from now on the Party Committees should, without exception, act in accordance with the Constitution. They should not supplant either the administration or--more important--the judiciary. The judicial bodies must not pass sentences without having ascertained and studied the relevant facts."[25]

As noted in chapter 3, the preponderance of Party influence was also said to curb the effectiveness of non-Party government officials. One League member called for "a system to guarantee that the non-Party people would be able to exercise their authority smoothly."[26] Another pointed out that the problem was not that DPG people did not hold offices, but that they could not "command the authority that normally goes with such an office."[27] It was frequently pointed out that non-Party officials were not given the information they needed to do their jobs.[28] People who held what might otherwise be significant posts sometimes had so many other positions that they were unable to execute their responsibilities. But the underlying problem, according to long-time Revolutionary Guomindang figure Cai Tingkai, was sectarian attitudes on the part of Communists.[29]

It also became apparent that tensions between Party and non-Party government people were not restricted to the central government but were manifested on the local level as well. According to Shao Lizi, "The general public attaches little importance to the PPCCs at the county level. The CP county committees . . . exercise much greater power."[30] Breakdowns in the United Front, Zhang Zhizhong (RG) indicated, were even more severe at lower levels than at higher ones. Indeed, there was blooming and contending during these weeks among the DPG chapters in many localities.[31]

With so many academics among their ranks, it is not surprising that DPG members would have many complaints about conditions in educational institutions. Perhaps the most outspoken academic was Chen Shiwei of Lanzhou University, the local Jiusan chairman. Chen insisted that "colleges do not need the leadership of a Communist Party Committee." At Jiusan forums and in radio broadcasts, he advocated that universities be run by the professors (though the Party might provide guidance in policy matters). At his university, he said, "vitality is low." He sharply attacked the university administration as incompetent and uncaring. The president was "rich in the idolatry of peasant revolution" but was a tyrant who did not want good professors because good professors could not be pushed around.[32] (See appendix 1 for additional discussion of the problems of academia.)

Of special interest are the views of three DPG leaders: Zhang Bojun (PW chairman and DL vice-chairman), Luo Longji (DL vice-chairman), and Zhang Naiqi (NCA vice-chairman). Zhang and Luo were even more important in the DL than their titles indicated. When the Communists reorganized the League in 1950, the top positions went to two docile elder statesmen, Zhang Lan and Shen Zhunru.[33] But Zhang Lan died in 1955. The subsequent leadership of elderly Shen was ineffective, and he virtually retired in March 1957. Thus, much of the responsibility for leading the League fell on the two younger men.

In addition to his democratic party posts, Zhang Bojun was minister of communications, vice-chairman of the CPPCC, and director of *Guangming Daily*. With his dual journalistic and

DPG positions, Zhang was in a unique position to take advantage of the situation when the Communists ostensibly let down their guard at the time of the May Forums. It now became clear that he hoped to see the DPGs become more independent, so that they could play a greater role in the nation's political life. As he later stated in one of his confessions. "I admired very much the Polish democratic parties after the October [1956] Incident because they managed to take action. I felt that [in China] the CP had exercised too tight control over the DPGs. This was particularly so with local organizations. Cadres were chosen by the Party on their behalf much to my dissatisfaction. When the CP advanced the line of long-term coexistence, it seemed to me that the democratic parties had a long way to go."[34]

Zhang was accused of causing a "malignant expansion" of both the League and the PW. He certainly had definite ideas as to what the future function of the non-Party groups should be. He shocked many by suggesting that the CPPCC be made into an "upper house," which would place the DPGs in a position to check the actions of the National People's Congress--and of the Communist Party.[35]

Luo Longji, an American- and British-educated political scientist and journalist, was made minister of timber industry in 1956. Since 1952 he had been a member of the administrative committee of *Guangming Daily*, which he sought to make an effective organ of the Democratic League. In 1957 he pointed to the problem intellectuals faced: They had to take orders from petty intellectuals of the proletarian class. Though professing support for the Communist Party, he argued that advanced intellectuals should be better utilized. "There are returned students from England who make their living as coolies, and returned students from the United States who run cigarette stalls."[36] Luo considered himself to be part of a "loyal opposition." Indeed, the regime trusted him enough to represent China at the World Peace Council in Ceylon in June 1957. Thus he was away when the anti-rightist campaign began, but he returned to face the heat.

Luo and Zhang would be attacked not only for their own views, but for leading a national anti-Party campaign. This, of course, they denied. As Zhang told a PW meeting in July: "In all honesty, I neither ordered nor hinted directly that fires should be lit."[37]

Zhang Naiqi, aside from being vice-chairman of the National Construction Association, was minister of food and had been a delegate to the National People's Congress. When, at the May Forums, he strongly urged his fellow participants to speak out fearlessly, no doubt many of them trusted his counsel. Zhang Naiqi himself struck at several mainstays of the Maoist political order. He denied the existence of the so-called dual character of the bourgeoisie, saying that the class was not basically different from the working class.[38] Even people who received fixed interest payments on investments were not necessarily exploiting but were simply the recipients of income-without-labor.[39] The whole concept of reeducating the bourgeoisie was challenged. Remolding, said Zhang, was "doctrinaire."[40] By implication, he was casting aside what even he had once said was Construction's raison d'être.[41] And in referring to that section of the masses of which the NCA was supposed to be the backbone, Zhang Naiqi was later claimed to have said: "Industrialists and merchants may love the country, but they absolutely will not love socialism."[42] Still, the Communists should not be paranoid. "In contending, one must not be afraid of one-sidedness; there are bound to be errors and shortcomings, but this is no cause for alarm."[43]

A word is in order on the activities of the DPG press during these weeks. These publications, which until now were more organs of the Communist Party than of the groups they ostensibly represented, faithfully printed the contentions of DPG spokespersons. This was especially true of *Guangming Daily*, but similar items were printed in other democratic party journals,[44] and to a somewhat lesser extent in standard Communist publications.[45]

Guangming Daily had been the organ of the Democratic League in the 1940s, but after 1949 it was obliged to "serve" all the DPGs (by supplying their members the Communist line). But on its staff were numerous DPG figures, including some quite independent-minded ones. Zhang Bojun, as noted, was its director. The editor at the time of the May Forums was Chu Anping.[46] Beginning in 1956, it was Chu Anping's policy to print all anti-Party statements without the usually required "analysis." And he appointed to the paper's staff several writers who were later condemned as rightists.[47] Chu and his associates sought to print more news of DPG activities, information about science and technology in capitalist countries, and accounts of the activities of various socialist bloc parties.[48] In addition, they were later accused of having printed news from foreign news agencies--again, without "analysis."[49]

Chu was skeptical about the idea that the democratic parties should aid the Communist Party in its rectification. They would do their "humble bit" but were too weak to do more: "The Party alone must consider the problems of how to ease the contradictions [between itself and non-Party people], how to harmonize relations between the Party and the masses, how the Party can . . . respect the position of non-Party people . . . how the Party can be more tolerant in its political arrangements, how it can rule by virtue, and how the nation's people, the talented ones or just the plain common people, can be properly placed."

In what would be an often-quoted statement, Chu said that the key to the whole problem "lies in the idea that 'the world [*tianxia*] belongs to the Party.' I think that a party leading a nation is not the same thing as a party owning a nation. . . . It is natural that the Party should want to remain strong and hold the key positions of power in government. But is it not too much that throughout the nation there must be a Party man in every unit, large or small, whether section of subsection; or that nothing, whether major or minor, can be done without a nod from a member of the Party?"[50]

It is not surprising that one of the first places the Communists turned to reshuffle personnel in the summer of 1957 was *Guangming Daily*.

The Aftermath

In the May 15 issue of the Revolutionary Guomindang journal *Unity* it was mentioned that, although no decision had been made on the matter, the RG Central Committee was "studying the question" of a rectification *within* the organization. Although it may be assumed that the Communists had already decided on such a rectification for the democratic parties, RG members were told that pending a decision on the matter members were still to concentrate their energies on aiding the CP in its own rectification.[51] Doubtless sensing something in the air, the Democratic League nervously held a high-level meeting on June 6 (see appendix 2), but it was too late to save the situation. Two days later, a *People's Daily* editorial ("What Is This

For?") signaled the end.

Each democratic party headquarters soon announced that the organization would undergo a rectification. From then on, nothing would happen spontaneously. Indeed, all activities of these groups during the summer and fall of 1957 were directed by the CP's United Front Department. After the close of the May Forums,[52] great pressure was applied to those among the DPG leadership who had demonstrated hostility toward the Communist Party. Of course, many DPG figures had remained quiet during this period; now it was their turn to apply group pressure on the rebels.

Zhang Bojun was among the first important DPG personages to recognize his errors publicly. His recantation came on June 19 and was immediately followed by similar statements from others. Zhang's tone was apologetic, but it is apparent that he had not been completely broken. "A person cannot guarantee that all his words are correct, even though he thinks they are right at the time he utters them. After criticism and reminders by other people, one can undertake a self-review to see what things are correct. This is of help to me." It was "possible" that he had said things antagonistic to the Party and had tended to undermine its leadership, but actually he had only meant to criticize people in lesser positions. Furthermore, according to a *People's Daily* paraphrase of Zhang's words,

> It is possible that the situation was not all that serious. He had spoken on the question of a Planning Board for politics, and planning is only a job of engineering and technical personnel, who did not hold leadership positions. He had also spoken on the question of procedure at meetings of the State Council. Perhaps in view of his political position it was possible that his words were rather vague. He would not defend himself. He would not say things which did not spring from the bottom of his heart.[53]

Two days later Zhang Bojun publicly turned against Chu Anping, and promised to keep *Guangming Daily* "on the road to socialism."[54] (Before having an opportunity to carry out this pledge, Zhang himself was relieved of his post on the paper.) The same day that he made his statement, he was severely criticized at a PW Central Committee meeting over which he presided.[55]

Zhang Naiqi made various statements in self-defense, including a spirited rebuttal to attacks on him.[56] He denied that he had ever been hostile to the Communist Party or to socialism.[57] Finally, on July 17, he did admit some ideological shortcomings, but he still insisted that he had never opposed Communism. His critics dug deep into his past, and he was mercilessly attacked for alleged "Hitlerite" support of the Guomindang. Over the next few years he would be dismissed from his various posts, including those in the National Construction Association and *Guangming Daily*.[58]

Meanwhile, a Democratic League rump group was maneuvering toward center stage. On June 13 this DL "Central Committee group" held a meeting at which rightists were attacked. According to *People's Daily*, "All the speakers criticized the mistaken views [that League leaders had been expressing]. . . . The speakers also unanimously demanded that the Central Committee of the League clarify its stand on such mistaken views and called upon the CC . . . to demarcate ideological boundaries from these mistaken views. Otherwise they would not be able to practice long-term coexistence with the Communist Party on the basis of socialism."[59] The self-criticisms of the various DL leaders were declared wanting.

The anti-rightist drive was quickly carried to the provincial level. Spokespersons for the various units in Beijing Municipality demanded that a meeting of the Central Committee's Standing Committee be held to clarify this situation, review the work of the League, and "improve its leadership." In Shanghai, the local League leader was condemned by the Municipal Committee. He had "consistently embraced views and carried out acts against the Communist Party." The New China News Agency published reports of similar activities in Tianjin, Xi'an, Guangzhou, Chengdu, Chongqing, Jilin, and Suzhou. All the reported meetings, including those in Beijing and Shanghai, took place within a two-day period.[60]

On June 18, Li Jishen condemned Revolutionary Guomindang figure Long Yun for the anti-Soviet views Long had expressed.[61] It is now known, however, that Long Yun had been only slightly ahead of his time. Primarily, he had questioned the appropriateness of China's alliance with the Soviet Union. He complained that Soviet loan terms had been too harsh, comparing unfavorably with what the United States charged its allies. The USSR, furthermore, should have shared the expenses of the Korean war. Factories in Manchuria should have not have been dismantled by Russian occupiers after World War II; the equipment should now be paid for.[62] On all this, Long Yun (and like-minded DPG figures[63]) would have the last laugh.

On June 21, the Jiusan announced the rectification that would take place within its ranks. Recruitment of new members was to be temporarily suspended.[64] Two days later a similar announcement was made by the Association for Promoting Democracy: Because of the rectification, there would be no further recruitment work for the time being.[65] Presumably the same plans were being laid in the larger DPGs, though only the case of the Revolutionary Guomindang can be satisfactorily documented. On July 6, the form that rectification in the RG would take was announced. The focus would be on the provincial and municipal levels. Chapters were to study the documents that would be coming down. No new members were to be recruited. Continued support for the CP rectification was to be given.[66]

In the August 8 issue of the Party journal *Study* an article by Xu Daohe appeared in which the author attempted to clarify the role and status of the democratic parties. He reaffirmed the principle of long-term coexistence and mutual supervision. But the DPGs were, after all, *bourgeois* parties. "To gloss over this distinction will lead to the erroneous idea of 'sharing the limelight' and 'standing on equal terms.'" The capitalist notion of the equality of the various parties was explicitly rejected.[67]

By August the democratic party rectifications were fully under way. There was a steady stream of announcements about the number of rightists in various parties. According to "preliminary statistics" cited by *Guangming Daily* on August 9, "more than 100 hard-core right wingers have been exposed among the members of the Revolutionary Guomindang."[68] On the fifteenth the New China News Agency reported that the Democratic League had been in great danger from rightists;[69] on the following day it reported that the Peasants' and Workers' Party had been "so heavily infiltrated by rightists of various descriptions that it was at the brink of paralysis and disintegration," largely because of Zhang Bojun's policy of "malignant expansion."[70]

There is, or course, much hyperbole in these charges. Nonetheless, it now became clear that all along there had been two factions within the various democratic parties, one that was independent-minded, and the other accommodative toward the Communist Party.[71] Often, sections were dominated by the dissidents. (In Wuhan, they controlled forty-two of seventy-three League chapters.[72]) Elsewhere, the accommodation faction may have been in the majority, but they remained deliberately quiet (perhaps on instructions) during the spring of 1957. In the summer, they were quiet no longer. These nonliberals attacked the liberals with fury. Examples were Wu Han's attacks on Luo Longji, and Wu Dakun's sharp criticism of Zhang Naiqi. Often, DPG lines were crossed, as when Hu Yuzhi of the Democratic League inveighed against the Peasants' and Workers' Party.[73] Likewise, one individual (one of my interview informants) recalled the past of one famous liberal, dredging up much petty dirt about the man, who was seen as having been a crypto-Nationalist, financially corrupt, a bigot, and a friend of warlords. Since the mid-1950s, the liberal had been stirring up anti-Communist sentiment, an activity claimed to be understandable in view of his alleged past. (By now this informant's views have softened somewhat, but he clings to the conviction that the liberal had been "arrogant.")

Inasmuch as the DPGs had swung so far to the "right," the Communist Party felt obliged to reconsider its attitude toward them. As a late-August *People's Daily* editorial pointed out,

> It has been proven that once they deviate from the Communist Party's leadership, the democratic parties are bound to lose their bearings, go in the direction opposed to socialism and lose the confidence of the people and the possibility for long-term coexistence. To rely on the leftists, unite and teach the middle-of-the-roaders to turn to the left, isolate the rightists and cause them to disintegrate, and to serve socialism truly under the leadership of the Communist Party--such is the sole correct line, and the way the DPGs must go. Only thus can there be long-term coexistence and mutual supervision.[74]

This explicitly leftist decree implied that the bourgeoisie's alleged progress (about which the leadership had been so proud) had simply not taken place, and that it could be fostered by the DPGs only if they were under the tightest CP control.

Exposures of democratic party errors continued through the

fall of 1957. In the September 12 issue of *Unity* it was announced that 20 (out of 132) members of the Revolutionary Guomindang's Central Committee had been rightists. And in the 24 provincial-level committees, almost one-fifth of the committeemen were rightists.[75] The next month a reported 2,199 rightists were exposed within the membership of the Democratic League, including leaders at all levels.[76] Eventually, approximately 6,600 League members (one-fifth the total) would be declared rightists or worse.[77]

During the latter part of September, five of the democratic parties held rectification work conferences in the capital.[78] Guidance was soon issued to local organs on carrying out the rectification, and now the chapters were instructed to take part.[79]

There is considerable uniformity in the events in the DPGs at this time, and all appears to have taken place under the hand of Li Weihan. It was Li who announced what issues should be taken up at local rectification meetings. Among his "Eight Points" was "the class basis of society and the two-faced character of the bourgeoisie." There were additional points to be discussed (other than the Eight) in parties comprising intellectuals--for example, "the intellectuals serving workers and peasants." And the NCA should discuss, aside from the Eight Points, seven additional ones, including "the advantages of the socialist system [to businessmen]."[80]

The confessions of high democratic party personages continued. On November 29, Chu Anping "admitted his guilt before the people and promised to repent and become a new man."[81] The next month Luo Longji made a partial confession, but he said he would not concede some charges even if it cost him his life.[82] Zhang Bojun made a written confession acceptable to the Party.[83]

In January 1958, many DPG people were dismissed from their government posts. Zhang Bojun, Zhang Naiqi, and Luo Longji all lost their positions as heads of ministries.[84] A complete list of those who no longer held positions of leadership within the DPGs was published; it included everyone who had taken advantage of Hundred Flowers to make serious criticisms of the Communist Party.[85]

How are the unorthodox developments that took place prior to June 1957 to be explained? Had the course of events been smoother, one might be inclined to conclude that Hundred Flowers was simply a Communist device for ferreting out the disloyal. But the shattering of the United Front was far more costly to the regime than was latent dissent. It has also been suggested that Hundred Flowers was a calculated easing of pressure, to insure that China did not see events similar to those that had shaken Eastern Europe the previous fall. But there never was a real danger of this in the PRC, and no one gave it a thought until Chinese developments had gone so far that it seemed necessary to reassert pressure.

The turn of events did not surprise everyone. Although the Hundred Flowers idea was supported by a broad spectrum of the leadership (from Mao to Deng Xiaoping), it was also opposed by some, notably the army's Chen Qitong, who as early as January complained that the attack on formalism was being twisted into an assault on the principle that culture must serve politics.[86]

Mao and his colleagues permitted the various segments of the bourgeoisie to take liberties because it was believed that the Party was politically strong, and that it could afford to invoke these policies. At the same time, it was realized that the concrete advantage cited by Mao Zedong might accrue from them. Most of the world's Communists see force and repression as necessary evils, and China has shown a distaste throughout its history for such methods. The Party hoped that its authority over the bourgeoisie, as over all social groups, rested on something more than these. After the disillusionment, it did not hesitate to reapply them. Although Mao Zedong lost considerable face, it was others who would have to pay the full cost of his mistake. Among them were members of democratic parties.

Mao's high-level critics, notably Liu Shaoqi and Peng Zhen, were quick to take advantage of the fact that they had correctly perceived that Hundred Flowers would be a disaster. Unlike Mao and even Deng Xiaoping, they may never have relished the idea of the Party being criticized by outsiders (though they did favor greater democracy *within* the CP).[87] The person they selected as spokesman for their cause was an interesting choice: Wu Han. Wu was deputy mayor under Peng, and a leading nonliberal DPG figure. (He was chairman of the Democratic League in Beijing Municipality.) Ostensibly, Wu's targets were his more liberal League colleagues, but, as is so often the case in Chinese politics, things were not as they seemed. The main proposition with which he took issue in 1957 was Luo Longji's proposal for a committee to reexamine the excesses of the campaigns against counterrevolutionaries of the early 1950s--a proposal that Mao seems to have been promoting for some time.[88] Wu complained that "Luo" (read: Mao) had shown "lack of trust" in Communist cadres. Likewise, press references to Zhang Bojun's comments regarding Party control of educational institutions appear to have been aimed at Mao Zedong.[89] By 1958, Wu's complaints were even more obviously directed at the Maoists. He questioned the quantity-over-quality principle in education and scholarship. He also objected to the rewriting of Chinese history and infusing it with anachronistic sociopolitical concepts. Chinese historians, said Wu, had an excellent tradition of writing history with fearless objectivity.[90]

Thus, on the heels of Hundred Flowers, lined up against Mao were the *apparatchiki* allied with survivors (nonliberals) among democratic party people. This unbalanced alliance would hold together for almost a decade. Only after 1965 would the Chairman have his revenge against both.

6
The DPGs in Eclipse, 1958-1978

The period between Hundred Flowers and the 1978 Party Plenum was a quiet one for the democratic parties, considering what the nation in general was enduring. Individual members had horrendous experiences, and the DPGs had little in the way of an institutional life. Surprisingly, however, the organizations did not die.

The Immortals

Not a great deal needs to be said about the first ten years.[1] It was a decade when the democratic parties were once again merely on the receiving end of Communist Party dictates. Beginning in 1959, a series of "meetings of immortals" were held. This term was perhaps intended to be ironic, the idea being that DPG members' sense of superiority had to be overcome. But the Party, in no little difficulty during these years of the failing Great Leap Forward, apparently did not want another brutal campaign, so those who had not already been declared rightists were dealt with more benignly. Officially, the "hundred flowers" were still to bloom, and democratic parties' meetings, we are told, were characterized by "gentle breezes and soft rains." Indeed, there was a brief but significant thaw, though only a few DPG people attempted to take advantage of it. One who did was the phenomenal Wu Han. But Wu's involvement in the DL (he was vice-chairman) seems largely unrelated to his political activities during the early 1960s. More significant was the fact that he belonged to the Communist Party,[2] as did most of the participants in this episode of "blooming and contending."[3]

One non-Communist leader who did try to make himself heard was Huang Yanpei of the NCA. In the early 1960s Huang set out to write his memoirs, considerable advice to the contrary notwithstanding. The pressures on him were such that he soon found he could not write the kind of historical analysis he wanted. In fact, the book was actually written by others, based loosely on what Huang dictated--this being the only way he was permitted to proceed. Much important information was omitted or deemphasized. During the Cultural Revolution this manuscript (which was not published until 1982) would be deemed one of Huang's "crimes."[4]

Nonetheless, by 1963 the groups seemed to be regaining some of their old luster. Now quite a few members who had been deemed rightists had that label removed. Furthermore, statements supportive of the democratic parties were forthcoming from people like Liu Shaoqi, Deng Xiaoping, Peng Zhen, and even leftist Kang Sheng.

The Cultural Revolution

Liu, Deng, and Peng were purged within a few years, however, and with them seemed to go any hope for the democratic parties. They were convenient targets for opportunists like Guo Moruo, who observed in 1965 that such united front organs had been adhering to the bourgeois line of "three reconciliations and one reeducation."[5] Radicals declared the DPGs to be reactionary, their leaders counterrevolutionary and enemy agents. On August 23, 1966, Red Guards sent the groups ultimatums, demanding that they all disband within three days. Within the specified time limit, the organizations announced that they were ceasing their activities.[6] A sign went up in front of the Beijing headquarters of the Democratic League claiming that the organization had been dissolved.[7] (This was not actually the case; perhaps the sign was only to deflect Red Guard wrath.) Sometimes democratic party leaders did not even put in the routine appearances at national festivities, but on other occasions they unaccountably would be trotted out. Presumably they were pawns of warring factions within the Communist Party. In general, their earlier support for nonradical CP figures was now backfiring.

The situation is exemplified by developments in Beijing Municipality. It will be recalled that Deputy Mayor Wu Han's attacks on other DPG people had been Mayor Peng Zhen's way of criticizing Mao Zedong. Thus, it is not surprising that Wu, a Democratic League luminary, should be caught up in the new tumult.[8] He had been a DL member since 1943. In the early 1950s he criticized himself for having a "supraclass viewpoint"[9] but thereafter was quieter. He probably joined the Communist Party around 1956, though this was long kept quiet. He did not criticize Mao's policies until after Hundred Flowers, when, as was noted in the previous chapter, he was Peng Zhen's surrogate. Then, Wu embarked on what now appears as a reckless undertaking. In 1959 he wrote a play called "Hai Rui Dismissed from Office," about an upright Ming dynasty official who had championed the rule of law and given land to the peasants--thus infringing on powerful interests, who drove him from office. The play was published in 1961 and then produced on stage. Audience reaction was favorable. But in 1965, leftists saw in it a plot to defeat the revolution. The opening salvo was penned by the as yet little-known Yao Wenyuan[10] (later of Gang of Four fame), and others soon chimed in. It was recalled that the play had been written during a time of economic difficulties, when many people advocated slowing down the Great Leap Forward. There had been calls for expansion of the unsocialist private plot system, and for reconsideration of some sentences that had been meted out to political offenders in the 1950s. So the significance of Hai Rui, who centuries before had redistributed the land and dispensed the law with impartiality, was clear. Furthermore, the fact that Wu's play emphasized the *dismissal* of Hai Rui led to the suspicion (doubtless justified) that the playwright's concern was actually the dismissal of Marshal Peng Dehuai, arch foe of the leftists.

As the Great Proletarian Cultural Revolution got under way, the hapless Wu Han ostensibly became the number one target. At first a few publications, particularly local Beijing organs, supported him, or at least limited their criticism to nonsensitive matters.[11] But the "debate" quickly turned into a nationwide campaign against Wu and the issues (and people) he represented. At the end of 1965 Wu issued a tepid self-criticism;[12] it did not satisfy his critics. Now, everyone had to be heard from, including many intellectuals, who must have had very mixed feelings. They were, after all, attacking one of their own, and some may have guessed that they would be next. But others doubtless enjoyed being able to attack the Party in this legitimate fashion (Wu was, after all, a member). Surely there were those who relished the opportunity to take revenge on one who had abandoned the liberals, especially those in the Democratic League, during Hundred Flowers.[13] In 1966 the criticism of Wu Han increased in intensity, but by now he had much company. Democratic party figures not only came under attack but were subject to much persecution over the ensuing years.

Another Beijing deputy mayor purged at this time was Wang Kunlun, a member (and future head) of the Revolutionary Guomindang. In 1966 Red Guards branded him a "counterrevolutionary element" and he lost his position as deputy mayor. He and his family remained in disgrace for about seven years.

Even progressive figures among democratic party leaders were fiercely attacked. One such personage was Shi Liang, who had led the campaign against rightists like Luo Longji in 1957. (One day, fully rehabilitated, she would head the Democratic League.) It was a time of both political and personal struggle, and Shi is said to have been targeted on instructions from Kang Sheng (now showing his true colors) and his wife, Cao Yi'ou.

And yet, it could have been much worse for the democratic parties, which at least survived. The reason that they did is doubtless that they were protected by China's two most powerful leaders. It is well known that Zhou Enlai stood by numerous liberals, even during the darkest days of the Cultural Revolution. On one occasion, the premier had one Revolutionary Guomindang man airlifted to Beijing just before the man's home was invaded by Red Guards. On another occasion, Zhou sent his personal secretary to stop a potentially violent "criticism meeting" against an elderly DPG figure.[14] What is more surprising is the fact that Mao Zedong himself came to the rescue of these institutions. In October 1966, as the country was erupting in chaos, the Chairman passed the word that the groups were "still necessary."[15] As a result, the democratic parties were not formally abolished.

Still, terror was in store for many members, for no national leaders were really in control of events. The people who should have been protecting the DPG people--those in charge of United Front matters--were in no position to do so. (Li Weihan was branded a "counterrevolutionary revisionist" in 1967; he would not be rehabilitated for eleven years.[16]) Thus, the suffering of democratic party people was often indescribable. Wu

Han was apparently beaten to near death in prison and then taken to a hospital where he died. Many other DPG people lost their lives in the ensuing political struggles. One was Mei Gongbin, a founder of the Revolutionary Guomindang.[17] "Capitalists" were also a favorite target of Red Guards, so one may assume that many members of the National Construction Association perished at their hands.

Nonetheless, it was rarely DPG affiliation per se that made people targets of such actions. Rather, it was simply that the organizations' constituencies (intellectuals, etc.) were under attack. (In all the verbal venom unleashed against Wu Han, his Democratic League affiliation is not known to have caused him a problem.) Indeed, DPG membership was by no means invariably detrimental to a member; even during the darkest days of the Cultural Revolution it could actually be helpful. In the early 1970s the democratic parties (in their then ghostly form) continued to offer benefits for members and their families. For example, a youth who had been sent down to a village might receive special treatment if he had DPG connections. If he had a parent well-placed in these organizations, the relationship signaled a "back stage" (*hou tai*) that might even enable the person to return to the city.[18]

Until Mao's death in 1976, however, the democratic parties existed almost solely on paper. There are no reports of meetings after 1966, although occasionally "respectable" DPG figures did surface. A few had government posts, and one even turned up on a provisional provincial revolutionary committee.[19] For most, however, the only safety lay in maintaining the lowest possible profile, waiting for the storm to blow over.

7
The DPGs Today

Mao Zedong was barely dead before efforts were undertaken to revive the democratic parties.[1] It is difficult to determine who was taking the initiative, but the signal came from Ye Jianying, who told an assembly of Zhejiang CPPCC delegates in 1977 that "The DPGs are here to stay."[2] Rebuilding these institutions would not be easy. Many members had been involuntarily scattered to some of the most forbidding parts of the country. Sometimes, teams were sent to search them out. Having been deemed "historical counterrevolutionaries" for so long, it is unsurprising that many were reluctant to rejoin causes that once got them in so much trouble.

The Communist Party had a tremendous task in rehabilitating the innocent, and it took many years.[3] The democratic parties played an important role in the process, with DPG-staffed committees studying and making recommendations in particular cases. Often rehabilitations were posthumous. Even in such a case, the "verdict reversal" is far from useless, for it may be devastating to be the child of someone "executed, imprisoned or subject to control." In some cases--even those the courts have refused to reconsider--teams largely composed of DPG leaders have the responsibility to reinvestigate old cases and make recommendations. Rehabilitation is far from automatic, and some cases hang in limbo long after the team has made a favorable recommendation.[4] But many cases have had happier endings.

From the point of view of the Democratic League, one important rehabilitation was the posthumous exoneration of playwright-historian Wu Han. In 1985, "Hai Rui Dismissed from Office" was produced under League auspices to mark the drama's twenty-fifth anniversary.[5]

Policies toward the democratic parties had now become as liberal as they ever had been, save for the 1956-57 experiment. Indeed, reminiscent of those years, the DPGs were urged to be forums for political criticism. The media tried to prime the public to expect this. Criticism of the Party was not only welcome, it was urgently sought. It all had a familiar ring, but to those who remembered 1957, *China Daily*[6] pointed out, "there it [is] in writing: a pledge that the Party will entertain fully criticisms from non-Party members, make earnest corrections, and strictly prohibit any retaliation against critics." The paper did not mention the wave of arrests in recent years of China's democratic activists, though there was an allusion to the current campaign against "spiritual pollution."[7] The DPGs were assured that the rectification would be aimed solely at the Communist Party, not at them, and that charges of lawlessness by Communists would be properly handled.[8]

The Communists bent every effort to improve the image of the democratic parties. During the winter of 1979-80, various Politburo figures spoke at DPG meetings. Deng Xiaoping told them: "The democratic parties are comprised of patriots who support socialism; they have now become political allies with the workers."[9] In particular, he looked forward to DPG specialists--those with professional expertise--aiding in China's modernization. He even specified that former Guomindang administrative and military personnel could make such contributions, as well as help improve relations with Taiwan.[10]

In the media, praise was heaped on the democratic parties, particularly their more obliging members.[11] At one well-publicized national meeting in 1985, United Front Department Chairman Yang Jingren congratulated the DPGs for "boosting morale in building socialist material and spiritual civilization." He looked forward to their "giving full scope to their political consultative and supervisory role by taking an active part in the state's political activities, continuing to contribute to the restucturing of the administrative system, opening to the outside world, and enlivening the domestic economy by gearing their work to socialist modernization, achieving a fundamental change for the better in the standard of social conduct by actively promoting socialist ethics, and broadening the patriotic united front by promoting contacts with compatriots in Taiwan, Hong Kong, Macao, and other countries." This was a good statement of the Communist Party's agenda for the DPGs.[12]

It was primarily the Communist Party members who needed to be persuaded of the wisdom of restoring the DPGs' "supervisory role." Despite efforts to upgrade the education level within the Party, the rank and file is still largely composed of people with limited knowledge and intellect.[13] These are the people who must bear the brunt of any DPG criticism. The leadership seems insensitive to the fact that if the intellectuals are really unleashed, Party members will be highly vulnerable. Nonetheless, in recent years the leadership has tried (with only limited success) to forge a new relationship between the Party and the intellectuals. An article in *Red Flag* insisted: "We cannot force all patriots to become socialists." Although it would be nice for everyone to be socialist, patriots of all sorts have a contribution to make to the motherland.[14] The press contained many articles urging better treatment and utilization of intellectuals in general, and democratic party people in particular.[15] It was pointed out that 70 percent of intellectual cadres were not Party members. "Cooperation with the DPGs and non-Party personages is of prime importance in the building of socialism with distinctive Chinese characteristics. . . . We must promote the best of these people to leadership positions and give them real power."[16]

In late 1983, the DPGs held congresses and ratified new constitutions. Although the texts were organized differently, there were no significant differences among the various documents, which generally contained similar wording, tailored to fit the peculiarities of each group. (See Appendix 3 for the text of the RG constitution.) Top Communist Party leaders spoke at these congresses, and Secretary General Hu Yaobang sent a statement. All were extremely deferential, referring to DPGs as "comrades in arms" and using terms like *gui-dang* and *bi-dang* ("your honorable party," "my humble Party"). But not all DPG people were impressed, some pointing out that the CP still contained some "bad elements" from the Cultural Revolution era.[17] Strong leadership would have been helpful at this time, but most of the DPG leaders were by now elderly. New replacements were found in 1983, but they were also too old to function effectively. Former minister of justice Shi Liang (born in the early years of the century[18]) died two years after being made chair of the Democratic League.[19] Her replacement, Acting Chairman Hu Yuzhi, likewise died in 1986, and was replaced by eighty-seven-year-old Chu Tunan.[20] An equally unviable nominee was Wang Kunlun as head of the Revolutionary Guomindang. In 1957 Wang had complained that the DPGs were "eyebrows," meaning useless appendages. At the conclusion of Hundred Flowers he joined in the attacks on his colleagues, particularly Long Yun and Zhang Bojun. Thus, he was able to remain out of trouble until the Cultural Revolution, when he went down with other Beijing Municipality leaders. But like his DL counterparts, Wang soon died.[21] The age of Zhou Jianren, new head of the Association for Promoting Democracy, is not known, but in 1983 the average age of the Association's chairman and various vice-chairmen was eighty-two.[22] Thus, whatever rejuvenation was taking place among the DPG rank and file, it was not happening at the top. This may suggest that the prospect of vigorous DPG leadership makes the CP nervous.

Also in 1983, the Central Socialist Academy was reopened in Beijing. Designed primarily to train DPG cadres and other united front functionaries, this institute had been founded in 1956 but did not function from 1966 until November 1983. It offers a two-year course of instruction, with students studying philosophy, political economy, modern Chinese history, theory and policy of the united front, CP-DPG relations, socialist democracy, law, and science and technology. The first graduating class had seventy-two students from all provincial-level units save one (presumably Tibet).[23] The academy may be the key to keeping the DPGs in line without resort to coercion. Not only does it insure that the cadres in charge of DPG affairs--the people who deal with the elected leaders of the local chapters--will be tuned in to Party requirements, it also should improve the intellectual level of DPG cadres so that they will not unnecessarily alienate DPG members.

Belonging to a Democratic Party

During the winter of 1984-85 I interviewed many DPG members in China in an effort to establish from talks with ordinary members what it means to belong to these groups. Although a few of my subjects were long-time members, I was especially interested in new recruits, on whom the future of these parties depends. Thus, my subjects were also relatively young, though all were at least in their mid-forties. A slight majority had joined their DPG since 1980; the others had joined in the 1940s and 1950s, that is, within a few years of the Communist victory over the Nationalists. Except for their "youth," my subjects were probably typical of the DPG universe. They ranged from liberal independent thinkers to active participants in the 1957-58 campaigns against their fellow intellectuals. Everyone was of the Han race.[24] All were males, which reflects the small number of women in the DPGs.[25]

People from four (out of the six or eight[26]) groups were interviewed. However, most of the subjects were members of the Democratic League, the largest DPG.

With one exception (the Revolutionary Guomindang[27]), the DPGs have all the characteristics of professional associations. In some cases, such as the National Construction Association, the distinction between a DPG and an explicitly professional society (e.g., the Federation of Industry and Commerce) is hazy, and functions, memberships and leaderships overlap.[28]

To become a member of most DPGs, one must be a fairly advanced "intellectual," which accounts for the paucity of members in their twenties and thirties. Although this membership criterion is well publicized, an effort was made to learn whether individual chapters adhere to it. In general they do, though some DPGs appeared to apply the criterion more rigidly than others. One of the strictest is the Jiusan, whose members must have associate professor or equivalent rank (but need not be affiliated with a university). Most work in the fields of science and technology, though my single Jiusan subject's field is literature.

The Democratic League appears to have somewhat looser recruitment practices. One subject said that all members of his chapter were professors and researchers, but another said that standards in his chapter were being lowered to accommodate younger members. (Because of the disruption in education during the Cultural Revolution, intellectuals tend to be either young or old.) At any rate, it was stated that standards are determined by the DPG, not by the CP.

Procedures for joining appear to be uniform among the democratic parties and are similar to (but less rigorous than) procedures for joining the Communist Party. First, one needs to be nominated by two members. Then the chapter and higher authorities must approve. The impression gained was that currently the Communist Party usually does not interfere in this process. At any rate, these are not organizations that would attract dissidents or troublemakers. Certain criteria must be adhered to. A recruit's political thought must be "standard" (*biaojun*), and sometimes people are rejected because of obstreperous political or other attitudes. One must support socialism and the four modernizations. But the more common reason for rejection is that one's academic level is inadequate. A few are turned away because they are too old and infirm. Three social groups are still out of bounds altogether: workers, peasants, and soldiers.

It is now rare for people to be dismissed from a democratic party. It has sometimes happened, especially in the National Construction Association for reasons of corruption. I learned of no recent dismissals for political reasons, but once these were common. For example, a highly committed democrat who was active in the Democratic League in the 1940s was squeezed out of the League by pro-Communists around 1950. He asserts that all other liberals were ousted at the same time. A distinguished scholar who refused to be remolded, the man subsequently endured two long prison sentences as a rightist. In 1984 he was invited to rejoin the League; he declined to do so. He still considers the organization a tool of the Communists and knows that he would not be allowed freedom to advance his political views.

Of particular interest are the motives that drove people to join a DPG. These vary widely. In paraphrase, here are some typical responses:

> I wanted to associate with like-minded people. I wanted to engage effectively in political work, and to have access to information that would make this possible. I refuse to be a bureaucrat.

> I don't admire the Communist Party. I would consider it shameful to become a member, which would mean becoming a member of the privileged elite. By joining a DPG I can participate in public affairs without belonging to the CP.

> I want to promote China's modernization, particularly in the field of education. Being a member of a sanctioned organization provides a measure of political protection.

> In 1957, when I was a teen-age deputy head of my Youth

League unit, I criticized some students for not attending classes, and I advocated certain reforms. As a result, I was declared a "rightist," a label that was not removed until 1979. I did not feel that my final vindication came until 1984, when I was allowed to join this DPG.

For many years I had had contacts with the Revolutionary Guomindang. What finally prompted me to join were the local chapter's plans to participate in commemoration activities for a national leader of the Republican era, Feng Yuxiang. A more general reason for joining was to help restore the good name of the Guomindang [the point being that although they came to be dominated by a reactionary faction, the Nationalists were not all bad].

I hated the Cultural Revolution; this is my way of reacting against it. I am a strong believer in moderation, especially in political tactics. I still oppose student demonstrations, even when the students' goals may be correct.

Had more subjects been interviewed, no doubt additional reasons would have been given, for it is obvious that people have a wide variety of personal as well as public-spirited reasons for entering a DPG.

Once a person joins, does membership in a democratic party live up to expectations? If so, are the benefits of belonging the same as those that led the individual to join in the first place? In general, the impression derived from my respondents was affirmative in both respects, but here one runs up against the pitfalls of my sampling technique. The interviews were all arranged through personal contacts. My universe of Chinese acquaintances comprises able, self-confident, self-directed individuals, and their friends are apt to be likewise. Under these circumstances, it is unlikely that I would be introduced to misfits or malcontents. Still, the stated benefits of DPG membership are worth recording. For example, one person said that "Now it is possible to convey opinions and make suggestions." Another remarked, "I respect the senior people in this DPG. Now I can see a lot of them and learn from them." And a third said, "We get to go on inspection tours. You need an organization to do things like this." (These excursions are more than a trivial advantage of DPG membership, as few Chinese have such opportunities. In 1985 the government set stringent limits on such out-of-town study tours, but the DPGs appeared to be exempted from the near-prohibition.[29])

Respondents were pressed regarding whether there are more material benefits to DPG membership. Some denied the existence of any perks. A more typical (composite) answer was: "They are not a major factor, but being a member of a democratic party may help in the assignment (*fen-pei*) of offspring. Our own work assignments are not affected until we become senior people; then our DPG superiors might put in a good word for us if they think we are entitled to promotion or other benefits, such as improved housing. This is especially true if one occupies a leadership role in the DPG chapter. Of course, if I should decide to change my work unit [normally impossible in China], my contacts throughout my DPG would be a big help."

A Democratic League national leader acknowledged that the organization is concerned about DL members' working conditions and livelihood, such as housing questions. "These little issues are important. The government is very big; it doesn't see these small problems." He had recently helped arrange a job change so that a husband and wife could live together. (When asked about the fairness of special treatment for those in his organization, he said that the League also helped people who were not DPG members.) This work of satisfying the needs of intellectuals enhances their "activism" so that they make greater contributions to the nation. But the persons interviewed, while often hoping that their organization could help them attain better housing, were usually still waiting.

Local authorities are inclined to provide these various forms of assistance (perhaps one should say favors) because cadres and Party members have a certain respect for the DPGs. They look on their members as "our kind of intellectuals." There is a certain irony here, for many intellectuals take a condescending attitude toward DPG people, considering them to be second-rate thinkers and opportunists. Though unfair in view of present realities, this is understandable in light of the backgrounds of these groups. (When the head of the Revolutionary Guomindang's Organization Department was told by a Hong Kong journalist that few people outside China had ever heard of his organization, he said he was not surprised. "Many people in China don't know [about us] either."[30])

Under all political systems, information is power. In China, with its largely closed information system, knowledge of policy developments and a sense of which way the political wind is blowing are tremendously important for an individual. The interviews revealed that democratic party channels are extremely useful for this purpose. Often it is simply a matter of learning of policy shifts before they are published. Such information might range from salary matters, to urban development plans, to foreign policy démarches. Sometimes information gleaned from DPG documents or discussions can show up quickly on a person's bottom line. For example, one respondent learned through his DPG channels that writers would be permitted to publish articles abroad and receive royalties for them. He acted quickly on the information, beating the pack. Another person, who happened also to be a member of the Communist Party, asserted that information often reached him through the League before it came down through Party channels.[31]

Members were questioned about any impact they have had on policy matters. The following, in paraphrase, are some typical answers:

> I feel that as a Democratic League member I am able to have a genuine input in the affairs of my university.

> I am a DL delegate to the Chinese People's Political Consultative Conference and can make suggestions at the proceedings. [Did not give any examples of having done so.]

> If there is something that needs to be discussed, I now can go directly to CP, which is a bit more inclined to listen if you are a DPG member.

> We are trying to obtain offices for local DL headquarters--the old place now being occupied by others. We have also lobbied to improve our DPG people's housing conditions, end spousal separation of League members, and return their possessions taken during the Cultural Revolution.

> We have urged that factories that pollute be moved from the city to the suburbs. We have pressed for diplomas to be granted where wrongly denied.

[A Democratic League national leader:] We regularly consult with DPG and nonparty personages on such subjects as: (1) Drafts of five-year plans. (2) Major economic projects. DPG people criticized the Baoshan steel project for two years before the suspension of construction; their suggestions were implemented when the project was resumed. The government also received much flak on the Bohai incident from the DPGs. (3) Reform of writing systems (characters). DPG people made input not only in the 1950s, but also the 1960s before the Cultural Revolution. The pressure to change from vertical to horizontal writing came from DPG people, who realized how important this would be for facilitating scientific and multilingual writing. (4) More money for education, especially for lower schools. (5) Middle-aged intellectuals' inadequate pay.

These final comments from the DL leader should perhaps be taken with some skepticism. One has to question whether China's leaders hear much that they do not want to hear.[32] Rank-and-file DPG members, though more able to be heard than others are, still have only limited political tools at their disposal. The leadership is well insulated from chapter meetings, the most common forums for airing views, and it controls the memberships of the other forums--the People's Political Consultative Conferences (PPCCs) at various lower levels.[33] Nonetheless, the CPPCC and PPCCs have work groups on many subjects,[34] and it is not inconceivable that word occasionally percolates up. Of course, (C)PPCC members have always had the right, individually or in groups, of interpellation (*ti'an*) at regular sessions, and the authorities are supposed to respond (*dafu*) with, at the very least, an explanation of why the complaint cannot be satisfied. But these bodies meet infrequently, and there were few real opportunities for this. Now, a new committee has been established to deal with suggestions between formal CPPCC sessions.[35] Some provinces hold bimonthly PPCC forums at which Communists are supposed to listen to non-Party views. "Mere lip service must not be paid; formalism will not suffice; practical results are essential."[36]

Occasionally, complaints and criticisms from democratic party members can be quite pointed. The best evidence for this comes not from my interview subjects, but from the media. In Guangxi, for example, DPG people told a 1984 Party forum that the reason intellectuals have such poor housing is that the good units go to bureaucrats. The area's education level was low because many teachers were unqualified. As for universities, they "have no personnel or financial powers" and should be given some autonomy. Also raised were the problems of political labels, professional ranks, food coupons, and salaries.[37] Similarly, Fujian DPG people pointed out how many Communists drag their feet when it comes to Deng Xiaoping's reforms. "Such cadres must be criticized, replaced, or dismissed."[38]

During the winter of 1984-85 there was a most impressive example of DPG involvement in policy making. The issue concerned the huge Three Gorges (Sanxia) Project for building a dam across the Yangzi River. The plan, which entails the creation of a whole new province, raised eyebrows abroad because of the ecological impact it could have. In China opposition was muted, but objections were nonetheless forthcoming.[39] The main opponents of the original plan were DPG members, notably engineers in the Jiusan Society. The organization succeeded in holding up the project for at least two years while it undertook a major feasibility study. The authorities have been making revisions in an attempt to satisfy the critics. The government would prefer not to press ahead with a risky project over DPG objections, as that could prove embarrassing if anything went wrong. There must be no repeat of Hundred Flowers, when the Communist Party ended up being embarrassed because it turned out that DPG people had correctly assessed Party errors.

It should be noted that the questions on which these people take an independent stand are those in which they have a legitimate concern--issues related to their area of expertise or to the situation of DPG members (intellectuals, etc.) per se. Only at considerable peril would one openly dissent regarding one of China's more far-reaching political controversies.

DPG regulations permit resignations. Respondents were asked whether anyone they knew had become bored and dropped out; all professed to know of no instance. Officials of the Shanghai Revolutionary Guomindang, asked what would happen if a person became inactive, stated that he or she would be kept on the rolls, but they professed to know of no such case. I learned of only one resignation from a DPG out of dissatisfaction--someone who "hated attending meetings." The most common reasons for becoming inactive or resigning are a person's moving to an area where there is no chapter,[40] and joining the army or the Communist Party.

The question of a person belonging to both a DPG and the Communist Party is a touchy one. In the past, some people in DPG national and provincial leadership roles were simultaneously CP members.[41] (Indeed, Communists or virtual Communists deliberately infiltrated the democratic parties in the late 1940s.) Since 1978, however, it has been the policy that DPG cadres should not belong to the Communist Party.

Normally, a Communist Party member may not now join a DPG. However, DPG members occasionally join the CP. For example, eight or ten RG members in Shanghai reportedly also belong to the Communist Party. DPG members might want to join the Communist Party so that they can play a greater role in their work units. Such people continue to belong to the DPG, but they typically become inactive. It would be awkward for such people to continue to attend DPG meetings because the question would arise as to which hat they wore.[42]

Before 1949 it was not uncommon for a person to belong to two DPGs, in part because the Democratic League was then an umbrella organization. After 1949, however, membership was usually permitted in only one party. Thus, one of my subjects had to choose between the Democratic League and the National Construction Association; he chose the latter.

DPGs are organized much like the Communist Party, with national headquarters in Beijing, intermediate (provincial or municipal) structures, and local chapters located in the workplace. By the end of 1983, the League had 2,300 chapters.[43] The size of the chapters examined ranged from fourteen to thirty people. Usually a work unit had only one chapter, but one university had four chapters of the Democratic League, each comprising people in a single specialty (science, Chinese history, foreign literature, and geography). The university also has a chapter of another DPG, and a few chapterless members of two other DPGs.

Chapter leaders were formerly selected by the higher levels. Now, elections are held every four or five years, and the chapters are more independent. So far as could be determined, the Communist Party does not send observers to elections. Approval by the DPG higher levels is still required, but most respondents said that this is routine. Chapters usually have a single leader, but sometimes there is a sort of board of directors. As an example of the first type of election, in one chapter there are usually three names on the ballot (with write-ins permitted), one of which is selected. As an example of the second setup, one Democratic League member reported that there are generally more candidates (e.g., nine) than the number of seats to be filled (seven). Incumbents are generally reelected. The winners always get 50 percent--there has not been a problem of winners only receiving a plurality. Most of the leaders are qualified. A few are not, or are elected just because they are famous. Such people, the respondent complained, do not have time to run the chapter properly.

DPG meetings are held approximately monthly, and there are occasional multichapter (e.g., citywide) meetings. Whereas in the 1950s attendance at meetings was sometimes low,[44] normally everyone attends chapter meetings, with exceptions accounted for by illness or scheduling conflicts. One person said that it is acceptable to skip a meeting if one asks in advance. There is often a predetermined topic of discussion. Mixed responses were received to the question of whether or not participation in discussion is de rigueur. Some said that it is perfectly all right to remain silent, but others said that this might be taken as lack of support for what was being undertaken or reported on.[45] Thus, not all of the old pressures associated with the DPGs have disappeared, but the obligatory mouthing of slogans seems to be a thing of the past.

Repeated inquiries were made regarding a chapter's relations with superior organs and with the Communist Party. Some chapters reported both to the higher DPG organ and to the local office of the CP United Front Department, but most respondents insisted that their chapter reported only within the DPG.[46] "The UFD provides guidance regarding national policies but has no role in internal the affairs of our DPG." Nonetheless, overriding questions are certainly settled by the Communists, such as the number of people representing each DPG in the CPPCC.[47]

As is the case with the salaries of the Communist Party's own cadres, those of the DPGs are set and paid by the state,[48] their salaries comprising the main expense of the provincial-level organs. If budget requests go much beyond salary needs, they are apt to be denied. Local chapters appear not to receive any funds from higher DPG organs, and they must be self-supporting. Members pay dues, typically one yuan (one percent of the monthly salary), but the assessments vary depending on the chapter's needs and the members' ability to pay. The lowest known dues are 0.10 yuan per month (for a Revolutionary Guomindang chapter member earning less than 100 yuan per month). The highest was 3 yuan (for people in certain chapters earning more than 100 yuan.) The National Construction Association is the best financed of the democratic parties, because its members are the highest paid. Other DPGs' chapters either get along without funding (often activities require little financing) or raise money in connection with their "contributions to the four modernizations."[49]

What the Democratic Parties Do

The Communists' underlying purpose in fostering these groups is to promote cultural and economic development. All subjects were asked about the ways their chapter was achieving these ends; the answers generally fell into two categories.

First, education is being promoted, primarily through the establishment of what may be called private schools. This is particularly true of Democratic League and Jiusan chapters. A city might have several of these schools, often providing part-time education using public school facilities after hours. There is at least one DPG college, a League-run institution in Hohhot specializing in foreign languages and technical subjects.[50] There are also mobile lecturers, who may even go to remote areas. Often the teachers are retired people (not necessarily DPG members), but the most dynamic elements behind the establishment of these institutions seem to be younger men who have recently joined the organization. Some schools provide adult education, some college preparation, and sometimes the students are children. Subjects range from technical topics to foreign language instruction. Tuition is modest, and the teachers receive only eight to twenty yuan per month. But the operation often provides the chapter with much of its income. Altogether the country has five hundred DPG-run educational institutions, with half a million students and graduates.[51] According to most reports, these schools function well and provide quality education. Some, however, encounter political and bureaucratic problems. A Guangxi DPG member complained: "The democratic parties encounter all kinds of restrictions in running schools. I hope the regional CP committee will seriously take care of education, open up more opportunities, and truly tap Guangxi's brain power."[52]

From the authorities' point of view, the advantage of DPG involvement in education is that schooling can be provided with no expenditures of public funds. Sometimes local authorities will provide classroom space or land for building schools, but it is generally up to the members of the DPGs to provide actual financing. The process usually begins with members providing seed money. Thereafter, tuition is supposed to make the institutions self-financing. (The average semester's fee is six yuan per regular class, and twenty yuan for a one-semester college course.)[53]

More lucrative for both the member and the organization is consulting work for enterprises. Small-scale entrepreneurs aside, China has not been a place where a person is apt to get far striking out on his own. With an organization like a DPG to legitimize and provide connections, however, one can do well. And since the DPGs are known as repositories of know-how, managers often seek help from them. Sometimes these consultants act pro bono, for example going to poor ethnic minority areas to help establish water works.[54] Among the fields are education, medicine, economic management, accounting, statistics, municipal works, and natural resource exploitation. By 1985, some one thousand enterprises in minority areas were being helped by DPGs.

Often, though, consulting can be financially attractive. Typically, a factory will pay two thousand yuan for a job, of which the individual receives 30 percent and the DPG the remainder. By mid-1985, the Revolutionary Guomindang had sixty-eight consulting services in economics, law, medicine, culture, education, science, and technology.[55] The Democratic League and National Construction Association doubtless had even more.

There are numerous miscellaneous ways that the DPGs promote political and economic development. For example, members may go around lecturing on the importance of the rule of law. Again, such activities can take place without the backing of such an organization, and other professional groups do provide similar services. But the DPGs appear to be making a genuine contribution in numerous ways.

Although a few DPG people would like these groups to be

parties in the internationally understood sense of the word, at present one of the less important aspects of these "parties" is fielding candidates in elections for government posts. Some local DPG organs do have members who run for various local offices (usually a PPCC, but sometimes administrative or local congress posts). Indeed, it is standard practice for a certain number of posts to be filled by such "democratic personages." Six percent of DPG members are People's Congress or PPCC representatives at the national or provincial level.[56] But no one tried to persuade me that the electoral candidates were picked by other than the Communist Party. And no one complained about this situation.

Respondents were asked whether they were in any way dissatisfied with their DPG, or whether things should be done that are not being done. The answers tended to be guarded, though one man was emphatic that no new projects should be undertaken. ("We've got too many things going on as it is!") Only one person gave a candid critique of his DPG--one of the smaller ones. He complained that, unlike the larger Democratic League and National Construction Association, his organization lacks infrastructure, and there are not enough cadres whose sole duty is DPG work. More tellingly, he had some words of criticism for his fellow members: There are not enough independent thinkers. People are cautious because of lingering fear. They remember the anti-rightist campaign and the Cultural Revolution. Thus, they are "too obedient."

Nonetheless, the DPGs are not the insipid and meaningless organizations they are widely perceived both in China and abroad. They have been undergoing considerable growth. Some chapters have become very large, and new chapters are being created. Many new provincial sections are being established. National infrastructures are being beefed up. The Democratic League is building a new headquarters in Beijing (sure to be less attractive than the present offices, which are located in the former home of a member of the imperial family.)

The increase in membership should rejuvenate the DPGs and lower the average age, though new recruits are generally over forty-five. In the case of the NCA, I was told that if recruits were too young they would have no value to the organization. They must be experienced before they can serve as consultants. Also, they are often obliged to serve without pay, in which case they must be retired and on pensions. Thus, the emphasis is on recruiting senior people.[57] But other DPGs have made headway in lowering the average age of members. In spite of its special membership requirements, 34 percent of the Jiusan were under fifty-six in 1983. Figures on the other DPGs are not available, but in the case of the League, three-quarters of new recruits are under fifty-five.[58]

Some members appear to be essentially "other-directed," having joined their DPG largely because their friends or peers were doing so. Others might be called more "inner-directed," that is, people with their own agenda, who see the DPG as a vehicle for accomplishing what they want to do. But even the other-directed people strike one as men of conscience, aware of a larger meaning of DPG membership, and hard working. (The Communist Party apparently agrees with this assessment, having given 30 percent of DPG members the prestigious label "model worker."[59])

The democratic parties have been slow to attempt to influence public opinion. Indeed, they lacked any means of doing so after 1957, when the *Guangming Daily* and their other newspapers were taken away from them.[60] In the spring of 1985, however, the Democratic League began publishing a new monthly magazine. Its name is *Cun yan*, which literally means "voice of the masses," but is generally rendered in English as either "Opinion" or "Tribune." The first issue (circulation sixty thousand) contained articles of major interest. Liberal political scientist Li Honglin wrote that China's power structure should be altered so that "everyone has a say."[61] Some articles dealt with more specific problems, such as the poor quality of education and problems of teachers. (It was noted that a dumpling peddler can earn more than a full professor.) There was also a piece about freedom in literary and artistic creativity, written by China's most distinguished journalist, Liu Binyan. *Cun yan* could be an important step for the democratic parties, but at this writing its significance cannot be assessed any confidence. After all, the publication is at least nominally committed to upholding the Four Basic Principles: supporting socialism, the democratic dictatorship, the Party, and its ideology.[62]

8
Conclusion

John Israel has remarked, in reference to intellectuals with ties to the Communist Party, that "in China, if you are not some kind of establishment intellectual, you are not a legitimate intellectual at all."[1] In a way, the democratic parties are the institutional counterparts of the establishment intellectuals. Like them, the DPGs are sanctioned and protected by the Party, because the Party needs them. Sometimes there is tension, and twice there has been a complete breakdown in the relationship. But usually the groups, like the establishment intellectuals, collaborate with the Party, knowing that if they do not they may lose everything.

There are other ways in which these groups reflect the larger Chinese political environment. Since 1949, China has been a land of ubiquitous "small groups" (*xiaozu*).[2] The purpose of these has usually been to enforce conformity, foster loyalty to the regime, and inspire hard work. Although the democratic parties can be seen as one more network of small groups, there are important distinctions. Though not true elites, they are elitist, not mass organizations. They are also voluntary associations, and enforcing discipline is no longer their responsibility. The DPGs have come a long way since the 1950s. From the Communist point of view, their main purpose used to be to turn fence-sitters into true believers. Hundred Flowers demonstrated that this had not succeeded, and that there were still many genuine liberals throughout the democratic party ranks. This is no longer a problem. Though many DPG members may be liberals by Chinese standards (little mindless mouthing of the CP line is heard from them), one no longer finds in them many of China's genuine free thinkers. Some things were indeed learned from the 1957 experience. Then, the DPGs were lightning rods, enabling the Communists to keep trouble under control, but they are no longer such. (Other organizations, of course, still serve this function.)

In certain ways, Hundred Flowers was similar to both Mao Zedong's quest for power in the late 1960s and Deng Xiaoping's in the late 1970s. In all three episodes can be seen the leaders' distrust of the Party bureaucracy. In Hundred Flowers, the liberals did Mao's bidding, but when they did their job too well, not only the cadres but Mao himself were humiliated, and Mao permitted cadres to turn against the liberals. Ten years later, many cadres were doubtless sorry, for now they could have used allies of any ilk in their life-and-death struggle against the radicals. The alliance between the *apparatachiki* and the democratic parties (now shorn of liberals) could not come into being until the late 1970s, when Deng Xiaoping had the DPGs revived. At the same time (but for more limited reasons), he also briefly gave free rein to the real democrats (those associated with the "democracy walls"[3]). These people were tolerated only to help him get the leftists under control. Once they had served their purpose, Deng dealt with the democrats just as the liberals had been treated in 1957: they were imprisoned. To some extent, the DPGs were revived for the same reason. The democratic dissidents could not be tolerated because their primary concern was China's *political* modernization, but it was a different story for the DPGs, who were only interested in Deng's limited, primarily economic modernization program.

These were democratic parties in name only. (I asked most of my interview subjects about the term. None tried to persuade me that the organizations were *democratic*. Most attributed to history the fact that the term is still used. One said frankly: The DPGs have "nothing to do with democracy.") Members have no real say in the affairs of their DPG, at least above the chapter level. The organizations do have a modicum of input into China's political process. At the very least, when people in officially sanctioned organizations become disaffected they can engage in foot dragging or "formalism." When the leadership tries to use them as sounding boards they can soft-peddle or damn with faint praise. They know that blatant insubordination would invite a repeat of 1957, but the Communist Party knows that such a scenario would also involve heavy costs. More important, the democratic parties' exclusivity may well render China less democratic than it would be if the DPGs did not exist, for they make the system of closed power more viable.[4] In short, the organizations signal a broadening of privilege at the top rather than at the bottom.

Nor are the DPGs *parties*. They are vestigial entities, and this generic term is an anachronism, like the word "state" as applied to what are really provinces in the United States. But just as the American states are real even if misnamed, the DPGs are not merely "flower vases," as is often asserted by their detractors. To be sure, window dressing is one of their functions.[5] But no one pretends that the Communist Party really shares power with non-Communist parties. China is a self-declared one-party system. Real parties, or at least successful ones, represent aggregations of interests, but in the case of the DPGs the interests have been *dis*aggregated. More important, the DPGs no longer seek political power. They comprise a loyal nonopposition.

Because they recruit from such limited populations, no democratic party can become large or powerful. To date, their maximum theoretical size is in the hundreds of thousands, not in the millions.[6] In a country of over a billion people, they are only dots on the social landscape. Even so, they are still a long way from living up to even their limited promise. They are not free of Communist control, and Party members still often think of the DPGs as extensions of the Party. (At one university a Department of Chinese Communist Party History subsumes DPG studies.) Indeed, the majority of members were chosen (or retained) because of their lack of propensity to rock the boat. This could conceivably change as the small echelon of new, more dynamic recruits grows, but for now the DPGs are top-heavy with cautious senior citizens. These are people who have learned, often the hard way, of the importance of subordination to the Communist Party. They cannot forget 1957, when the press was warning that the very survival of the DPGs depended on their slavish subservience to the Communist Party in general and to leftism in particular.[7] Small wonder that today the average DPG veteran is a timid soul.

Although it is not their main function, the democratic parties serve as legitimizing agents for the regime. They do this in several ways. First, they make it possible for the PRC to be presented to the world as a multiparty democracy. However, to be impressed a foreigner must be both well informed (to know of them at all) and naive--a rare combination.[8] In a very different way, the DPGs have a legitimizing function within China, for they serve as a reminder of the good old days--the 1940s and 1950s when the Communists, increasingly supported by the DPGs, won the civil war and then improved the human condition in the country. Finally, they are frequently trotted out to portray broad-based support for the government's domestic and international policies.[9]

There are also times, however, when the democratic parties function very differently from this. Although the evidence is scant, occasionally they seem to become drawn into the factional politics of the Communist Party. For example, in 1986 the League was pressed into service in the drive to promote "socialist ethics." The memory of the Foolish Old Man Who Moved Mountains (harking back to a Maoist theme) was invoked.[10] Also, in the controversy over the Three Gorges dam (see chapter 7), there have been indications that in opposing the project the Jiusan Society was acting at the behest of the leaders of Sichuan (a province standing to lose a lot if Three Gorges goes forward). But politics of this nature is probably atypical, and in the Chinese context it is not particularly legitimate.

More legitimate (and legitimizing of the regime) is the fact that the democratic parties fit well with Chinese concepts of how a people should be organized. Society is seen as divided into distinct sectors--peasants, workers, intellectuals, etc. Each group is entitled to "representation" in the government. Each group is a component in a corporate state. For the Chinese Nationalists, who flirted with fascist ideology in the 1930s, this notion has always served as an important organizing principle,[11] though one that was only achieved after they established themselves on Taiwan.[12] Under this system, intellectuals, scientific personnel, business managers, and other professionals each have their organizations, but these exist within the context of authoritarian control. In Western political theory this is called corporatism. The assumption is that "man should not be politically articulate as a *citizen*, but only as a worker, entrepreneur, farmer, doctor, or lawyer; general political problems are assumed to be too complicated for the mass of the people, who are only expected to understand issues that bear directly on their professional or vocational work."[13] Larger political issues must be resolved by the small group of national leaders who are able to understand them. There are strains of Plato, Burke, and Hegel in all of this. For the Nationalists, the authority to control the corporate labyrinth has been the *state*. In 1929 they abandoned the Leninist notion that the (Nationalist) Party should serve this role. But the Communists held to the principle of Party control. This explains why the United Front Department has the supervisory role, even though DPG cadres are paid by the state. The resulting arrangement nearly fits Philippe Schmitter's definition of corporatism as "a system of interest representation in which the constituent units are organized into a limited number of singular compulsory, noncompetitive, hierarchically ordered and functionally differentiated categories, recognized or licensed (if not created) by the state and granted a deliberate representational monopoly within their respective categories in exchange for observing certain control on their selection of leaders and articulation of demands and supports."[14]

For our purposes, the only problem with this definition is the "compulsory" attribute. But even though the DPGs are voluntary associations, so are many associations in corporate states--at least in the sense that if one is willing to pay the price one can opt not to join.

It is also noteworthy that Schmitter at the outset equates corporatism with interest representation. Obviously, one is not talking about anything like interest groups in democratic societies. Indeed, Joseph Fewsmith, in his account of Shanghai politics before 1930, observes that "authoritarian regimes, it seems, have a compelling interest in establishing hierarchical, noncompetitive interest associations that *do not work*."[15] In an open society there are more effective means of promoting interests than simply forming officially sanctioned and curbed organizations. Still, the causes the DPGs serve are not entirely removed from what political scientists mean by "interest"-- namely, the desire to affect the allocation of social values. The fact that these values tend to coincide with those of the nation's rulers does not make the term "interest groups" any less applicable, whatever the implications for democracy. But the time has long passed when China can be thought of as totalitarian, or even characterized by simple "two-line struggles." Modernization means complexity, and with technocracy replacing one-man rule, the system offers many pressure points. In the political process, there are winners (not solely Communist Party members), losers (not solely non-Party people), and those like the DPGs who win less than some but more than most.

To be sure, the DPGs do not function like U.S. lobbies. Nonetheless, like all interest groups, they work collectively to gain political "goods" for their members and (to a lesser extent) for similarly situated individuals outside the group. The causes may be either self-serving (e.g., better living conditions) or public-spirited (improving the administration of justice) or a combination of the two (establishing private schools, consulting services, etc.). Such dual benefits appear to be the prevailing situation. When, for example, the Association for Promoting Democracy presses for improvements in public schools, it is acting in the interests both of its members (teachers) and of society.[16] Interest groups are a plus for any society provided that there are checks against their unbridled influence--certainly not a problem in the case of the DPGs.

Some might argue that the democratic parties are not very modern organizations. First, they are characterized by a degree of functional diffuseness. Aside from being a special kind of political party, they are also professional associations, interest groups, philanthropic societies, and sometimes even ombudsmen involved in reversing verdicts and influencing the dispersal of perquisites. Second, they are ascriptive in the sense that a person must belong to a certain social group to be considered for membership. Indeed, one must be recommended by a member before being allowed to apply to join.

The DPGs, however, are much more modern organizations than this analysis would suggest. The various functions do not represent an arbitrary conglomeration; they fit neatly together, just as do the various functions of a typical Western corporation. As for being ascriptive, the various constituencies from which they recruit are all nonascriptively defined (except for the problem of low female representation). But most important, these are different from traditional Chinese organizations. There appears to be little clientelism involved.[17] None of the interviewees indicated that a personal relationship with a chapter leader was particularly important to him. Recruiting is not done within established social networks, and relationships are not sustained on the basis of gifts or personal services. Factionalism, the hallmark of Chinese politics, no doubt exists, but none came to light during the investigation for this book. The blatant patron-client ties that characterize the relationships of many of China's establishment intellectuals with national leaders seem not to exist in the DPGs. The only real divisions--those manifested in 1957--were purely political and issue-oriented. Thus, by Chinese standards, the democratic parties must be seen as modern organizations. They are not made up of old-fashioned factions, based on patrimony and bent on destroying each other. They are not clandestine societies whose political work requires secrecy. Rather, they are groups of people with a common ethos and political purpose, and similar or at least compatible interests. They are one more indication that while China is not a democracy, neither is it a totalitarian state.

The future of the democratic parties is now completely tied to that of the Communist Party. The operative slogan governing CP-DPG relations in 1986 is: "Treat each other with respect; stick together through thick and thin"--or, more literally, "through honor and disgrace."[18] Presumably, "disgrace" refers to the past; "honor," to the future. Nonetheless, there is an uneasiness about the slogan, and about the whole DPG-CP relationship. Doubtless many democratic party members wish their fortunes were not so closely tied to the uncertain honor of the Communist Party. Of course, if the regime's political and economic reforms succeed, all will benefit. One of those reforms involves rejuvenation of China's political infrastructure. For the DPGs, this is crucial. At this writing, the average age of the 177 provincial committee chairs is seventy-five. But 400 younger people are slated to be promoted to leadership positions in the provincial sections and national organizations.[19] This is expected to mean two or three new leaders under fifty-five on every central and provincial committee.[20] This should enhance the institutional health of the organizations, which already have surprising vitality. Their members now number 160,000.[21] More than half of the DPG members were recruited in the post-Mao period, and soon there will be little institutional memory of the unhappy Mao-era history. Thus, a decline of institutional timidity can be expected, which should enable the organizations to command greater respect. Indeed, the DPGs are now attracting distinguished people. About half of the 396 members of the general assembly of the prestigious Chinese Academy of Sciences belong to DPGs, including CAS President Lu Jiaxi. But to date, the DPGs--like the Communist

Party--do not stand as high in the public regard as they did in the 1950s.[22] Both have a long way to go to regain their prestige. If the Communist Party does not succeed, neither will the democratic parties.

Appendix 1
The Democratic League's 1957 Academic Program

The following is a translation of a document that appeared in Guangming Daily, *June 9, 1957, pp. 1 and 3, under the title "Specific Proposals Regarding Science: Program of the Temporary Study Group on Scientific Planning." The word "science" includes the social sciences, and indeed it is these politically sensitive fields that were primarily at issue.*

The program was dated June 1957. Headings and some numbers of subparagraphs have been deleted.

A. 1. We propose that, with few exceptions, able [social and physical] scientists be largely exempted from administrative work. This should be especially true for those over the age of sixty, who should be passing their knowledge on to the next generation. All scientists should have a period each year when they are entirely free to engage in research. . . .

2. They should have enough research assistance and administrative staff so that they can work efficiently.

We propose that scientists competent to lead scientific research (such as those in the Academy of Sciences) be provided with suitable assistance of their choice.

3. Currently, there are also problems in matters of housing, reference works, equipment, chemicals for experiments, materials for tests and specimens, etc. Equipment is still primitive.

The short-fall should be filled as soon as possible.

4. The main barrier in research is the severe and rigid security system. Except where military and diplomatic [secrets are involved], there should be no secrecy in areas of professors' expertise.

5. There are cases both of scientists without money and of money without personnel. . . . We propose the establishment of special funds in universities so that research plans will not fail for lack of funds.

C. . . . 2. Attitudes toward the social sciences must be changed before these fields can be developed. Since Liberation [1949], certain fields have ceased to be independent subjects. A number of sociologists, political scientists, and legal scholars have changed their professions. Some subjects have been dropped simply because they do not appear in the Soviet roster. No longer emphasized are subjects like capitalist countries, political systems, and international relations and international law--all once deemed essential. We consider this approach inappropriate. Traditional social sciences should be reformed, not abolished. Where appropriate, steps should be taken to reinstate such subjects. . . .

3. When it comes to economics and finance, people merely follow the dictates of those in charge of government departments, publicizing rather than elaborating on [the current line]. This is not good enough. In the spirit of searching for truth (*qiushi*), social scientists should conduct investigation and research, and on the basis of this submit proposals on government policies and statutes.

E. Promotions in universities and schools, and the selection of graduate students to study abroad, have been too much influenced by political considerations. Hereafter, we believe that as much weight should be placed on professional competence as on politics. The problem of unemployed scientists and inappropriately employed scientists should be corrected. . . . We support the decision of the State Council on the examination and selection of students going abroad for advanced studies.

Appendix 2
Emergency Democratic League Conference, 1957

The following describes a high-level Democratic League meeting held just as the Communist Party was preparing its campaign against right-leaning DPG people in 1957. This is from an article by Min Ganghou, a member of the DL Standing Committee. The piece appeared in People's Daily *on July 4, 1957, p. 2. It is particularly interesting for what it reveals about the attitude of DPG leaders toward student dissidents. The students are respected, but these "establishment intellectuals" are nervous about where the students' activities might lead. Interviews conducted in 1984-85 found similar ambivalent attitudes toward student demonstrations.*

On June 6 [1957] at 10 a.m., Zhang Bojun, first deputy chairman of the Democratic League and minister of communications, invited several well-known scholars to an emergency conference in the Culture Club of the CPPCC to discuss the present situation and to take action. Attending were Zeng Zhaolun, Qian Weichang, Fei Xiaotong, Tao Darong, Wu Qingchao, Huang Yaomian, and Ye Duyi, director of the League's general office. Also, Minister of Justice Shi Liang, Hu Yuzhi (who withdrew from the conference), Jin Ruonian, and I were invited to attend. Probably [we had been invited so that we could be] enlightened about the situation. . . .

Zhang Bojun said that the situation in the schools [a reference to recent student protests] was quite serious. He asked us to consider what the DL should do in connection with the movement.

Sociologist Fei Xiaotong [member of the DL Central Committee] was the first to express his views. He said that the university students had been turned loose and they had gotten all worked up. Judging from the problems which had come to light during the [rectification] campaign, it appeared that the situation was indeed serious. He had heard that two Beijing University students claimed that during the anti-counterrevolutionaries campaign they had been wronged. When they told their story, some listeners wept. "Things are too grim," Fei said. "We intellectuals cannot tolerate such [mistreatment]. My feelings have changed, and I sympathize with the students. But once they are aroused, the situation is likely to deteriorate. The students are looking for leaders. If teachers join in, though, there will be big trouble. Of course, it can all easily be repressed--three million soldiers could do the job. But in that event, public support for the Communist Party would evaporate, and the masses would no longer respect it." He added that the problem was the system. Non-Party people had no authority, and Party and Youth League members had the authority and usurped power. "I think it is not a question of individuals but rather the system. I have declared that I won't join the CP, that's my attitude." Qian Weichang interjected: "I certainly won't join the Party." [Fei continued:] "Some people say, 'Without Party backing I'm nothing.' I disagree. Let us see whether the people support me or not in an election campaign."

Deputy Minister of Higher Education Zeng Zhaolun spoke next. [In April, Zeng had headed up the Temporary Study Group on Scientific Planning, which was responsible for drawing up the program outlined in Appendix 1.] He said: "The students have lots of problems, and they can't take any more. Once the students are in the streets, citizens will gather and the situation will deteriorate. After all, the public also is dissatisfied with the Party." He said that in the early 1950s, after so much chaos, students had wanted to settle down and study. Party prestige was high then, and there were several years of calm. Now things had changed. The Party had alienated the people. Thus, with the impact of the Polish and Hungarian incidents, the situation had reached an impasse. Things in China were much as they had been in Poland before the Eighth Plenum of the Polish Workers Party. As far as the present rectification campaign was concerned, it was possible that the Chinese Communist Party had misjudged the situation. Perhaps it had thought that, though higher intellectuals were a problem, young students were not a problem. It turned out to the contrary. There had been trouble at Xi'an's Communications University. In Shanghai, the situation might be even more serious than in Beijing.

According to Qian Weichang, vice president of Qinghua University [and DL Central Committee member], it was especially noteworthy that the students were searching for leadership. If teachers took the lead, there might be trouble. "Some students' parents wrote asking me to dissuade their children from making trouble. I tried, but the students were determined. It's like the eve of the May Fourth [1919] movement. They would not heed their parents' advice, just as when we were students we paid no attention to our parents. The students . . . are hoping that we will come forward and speak to them on their behalf. But it is difficult to speak for them. Some in Qinghua University have suggested that the president should resign and I should replace him. . . ."

Zhang Bojun expressed appreciation for these statements. "Students from a Hankou school under the Ministry of Communications are going to present a petition. Elsewhere, there are student strikes. The situation is serious. . . . If students take to the streets and the general public follows them, things will get out of control. . . .

"I favor a major expansion of the democratic parties and groups. At least one or two million people should be recruited. . . . The DPGs should be established down to the county level, for only by so doing can they fulfill their supervisory function.

"During the current rectification movement, non-Party people are supposed to express opinions. I believe that the venerable Mao Zedong knew what to expect: The democratic parties always offer criticism in a polite manner. But the estimates were incomplete. It was not realized that the Party could have committed so many errors. More problems than expected have been pointed out. Truly, our task has been 'overfulfilled.' Now the CP has a dilemma; they cannot advance, but they cannot retreat. The Democratic League has the responsibility of helping the Party find a way out of its quandary."

Appendix 3
A DPG Constitution: The Revolutionary Guomindang (1983)

In late 1983 the democratic parties adopted new constitutions. The following is a sample. The constitution of the Revolutionary Guomindang, adopted in Beijing by the RG's Sixth National Congress on December 30, 1983, has been selected because of the senior position of that group. It is the oldest DPG and is considered the legitimate incarnation of the party that ruled China during the Republican period. It plays a central role in Beijing's appeals to the Nationalists on Taiwan. The constitutions of the other DPGs are similar except for portions necessarily tailored to match the character of the memberships.

Translation by New China News Agency, December 30, 1983, JPRS-CPS-84-009. For comparative purposes, the constitution of the Association for Promoting Democracy may be consulted. An NCNA translation appears in JPRS 85001.

General Program

The Revolutionary Committee of the Chinese Guomindang [Revolutionary Guomindang] is a democratic party of the patriotic United Front under the leadership of the Communist Party of China; it is a component unit of the Chinese People's Political Consultative Conference; it is a political alliance of some socialist working people and patriots who support socialism and with whom it has contacts; it is a political party serving socialism.

In the 1911 Revolution under Dr. Sun Yat-sen's leadership, a feudal monarchy was overthrown and a republic was founded. Our historical mission of opposing imperialism and feudalism, however, was not accomplished then. With the Chinese Communists' assistance, Dr. Sun Yat-sen in 1924 promulgated three major policies: allying with the Soviet Union, allying with the Communists, and assisting the peasants and workers; developed the old Three Principles of the People into the new Three

Principles of the People; reorganized the Chinese Guomindang [GMD]; achieved the first GMD-CP cooperation; and promoted the national and democratic movement in China. After Dr. Sun Yat-sen passed away, the democratic faction and other patriotic democrats in the Guomindang inherited his patriotism and revolutionary spirit of constant progress, gradually developed and amalgamated in the course of the Chinese people's democratic revolution and organized the Revolutionary Committee of the Chinese Guomindang. Under the guidance of the United Front policy of the Communist Party of China, this party contributed to the overthrow of the reactionary rule of imperialism, feudalism, and bureaucratic capitalism and to the founding of the People's Republic of China. During the period of socialist transformation this party, under the leadership of the Communist Party of China, played a positive role in the struggle to consolidate the people's democratic dictatorship, promote socialism, and oppose internal and external enemies, and was tempered and tested in the struggle. In the days to come, this party will continue its long and firm cooperation with the Communist Party of China and advance along with it.

The Revolutionary Committee of the Chinese Guomindang supports the leadership of the Communist Party of China. Under the guidance of Marxism-Leninism-Mao Zedong Thought, it upholds the socialist system and the people's democratic dictatorship, abides by the Constitution of the People's Republic of China as the basic norm for all its activities, and implements the Constitution of the Chinese People's Political Consultative Conference.

China has entered a new period in which its central task is socialist modernization. The general task of the new period is to work hard on the basis of self-reliance to modernize industry, agriculture, national defense, and science and technology on a step-by-step basis to turn China into a socialist country with a high level of culture and democracy. In the 1980s, the Chinese people's three major tasks are to step up the socialist modernization, to strive for the reunification of the motherland with Taiwan, and to oppose hegemony and defend world peace. The Revolutionary Committee of the Chinese Guomindang will continue to join the people throughout the country in striving to accomplish the general task of the new period and the three major tasks of the 1980s, quadruple the gross annual value of industrial and agricultural production by the end of the century, achieve the great solidarity and great unity of the Chinese nation, and revitalize China.

Following the policy of "long-term coexistence and mutual supervision" and the principle of "treating each other with all sincerity and sharing weal and woe" pursued by the Communist Party of China in dealing with various democratic parties, this party will actively participate in the political life of the state, defend the political situation of stability and unity, develop socialist democracy, improve the socialist legal system, hold political consultations on important national policies and on important questions concerning the people's well-being, and play a democratic supervisory role in the course of making suggestions and criticism. This party will work independently within the limits on the rights and duties as prescribed by the Constitution of the People's Republic of China.

The central task of this party is to serve the socialist modernization. It will bring into full play the socialist initiative and creativeness of all its members and the personalities with whom it has contacts to contribute to promoting the building of socialist material civilization and spiritual civilization and to the building of socialism with Chinese characteristics.

The emphasis of this party's work is on promoting the reunification of the motherland. It will propagate and implement the Chinese Communist Party's policies of returning Taiwan to the embrace of the motherland--to achieving a peaceful reunification. It will step up contacts with the Guomindang military and political personnel and their relatives in Taiwan, Hong Kong, Macao, and foreign countries, and unite with patriots who support the reunification of the motherland to strive to reunify the motherland.

The Revolutionary Committee of the Chinese Guomindang will recruit new members primarily from among former Guomindang personnel and from among the personalities of the middle and upper strata who have had relations with the Guomindang in the past. It will pay attention to recruiting representative middle-aged persons; it will establish its organizations primarily in large and medium-sized cities. Adhering to the principle of integrating consolidation with development it will develop itself in the course of its work, and it will develop itself to facilitate its work.

This party will urge and help its members and persons with whom it has contacts diligently to study Marxism-Leninism-

Mao Zedong Thought, raise their patriotism and socialist consciousness, and remold their subjective world while changing the objective world.

The organizations of this party at all levels should adhere to democratic centralism, exercise collective leadership, practice a division of labor with individual responsibility, display democracy, strengthen unity, persistently seek truth from facts, uphold a mass line, constantly reform themselves, have the courage to blaze new trails, and strive to create a new situation in the work of this party.

Chapter I: Membership

<u>Article 1</u>. Any citizen of the People's Republic of China who accepts this party's constitution and meets the requirements of its membership will become a member of this party after completing the induction formalities.

<u>Article 2. The Duties of Party Members</u>. 1. To uphold the four basic principles, abide by the constitution and other laws of the People's Republic of China, implement the line, principles, and policies of the Communist Party of China, and guard state secrets;

2. To contribute actively to building a high degree of socialist material civilization and spiritual civilization and to reunifying the motherland with Taiwan;

3. To observe the constitution of this party, implement its resolutions, pay membership dues, fulfill actively any task assigned by the party, take part in party activities, and conduct criticism and self-criticism;

4. To study hard Marxism-Leninism-Mao Zedong Thought, study the line, principles, and policies of the Communist Party of China, and learn scientific, technological, and general knowledge to enhance ideological and political understanding and improve their ability to serve the motherland.

<u>Article 3. The Rights of Party Members</u>. 1. To participate in pertinent meetings and activities of this party and make suggestions regarding its work;

2. To vote, elect, and stand for election;

3. To criticize any organization or member of this party;

4. In case of disagreement with a party decision, to make

reservations and present their views to the party's leading organ, provided that they carry out the decision before it is changed;

5. To exercise the right of self-defense at meetings held by the party organizations to decide on disciplinary measures to be taken against them; in case of disagreement with the decision, to demand a reconsideration or to appeal to higher levels of the party.

Article 4. An applicant for party membership must be recommended by two members of this party, and file an application form according to prescribed procedures. The application must be accepted by a general membership meeting of the party chapter concerned, approved by the next higher party organization, and then be submitted through proper channels to the Central Committee for the record.

The Central Committee and the provincial, autonomous region, and municipal party committees may recruit new party members directly if they deem it necessary to do so.

Article 5. If a party member wants to withdraw from the party he should formally file a withdrawal request with the party organization to which he belongs. The chapter concerned shall approve his request at a general membership meeting, report the matter to the next higher party organization to have his name removed from the party rolls, and also report it through proper channels to the Central Committee for the record.

Article 6. When a party member is moving from one place to another, he must go through the procedures for the transfer of organization affiliation.

Article 7. Party organizations at all levels should properly reward and commend party members if they display good performance in their work.

Article 8. Party members who violate the PRC Constitution, the state laws and policies, or the constitution and discipline of this party should be given warnings or serious warnings, suspended or removed from party posts, placed on probation within the party, or expelled from the party according to the seriousness of their acts. The period of probation within the party should not exceed two years.

Article 9. In general, disciplinary measures against party members should be discussed in grass-roots units and reported to the higher organizations of the provinces, autonomous regions, or municipalities concerned. All disciplinary measures against party members should be reported to the Central Committee for the record.

Disciplinary measures against members of local committees at various levels and against advisers to organizations of provinces, autonomous regions, municipalities, and cities with districts should be decided upon by the committees at the same level and reported to the Central Committee for approval.

Disciplinary measures against members and alternate members of the Central Committee, advisers at the central level, and members of the Solidarity Committee should be decided upon by the Central Committee.

Chapter II: Organizational System

Article 10. This party is organized according to the principle of democratic centralism.

1. Individuals are subordinate to the organization, the minority is subordinate to the majority, the lower organizations are subordinate to the higher organizations, and all local organizations are subordinate to the Central Committee;
2. Leading bodies at all levels are elected;
3. Committees at all levels are responsible to and report their work to the congresses or general membership meetings at the same level and the organizations at the next higher level;
4. Organizations at all levels implement the principle of collective leadership combined with individual responsibilities based on division of labor.

Article 11. The party's supreme leading body is the National Congress and, when it is not in session, the Central Committee.

The leading body of a local organization is the congress or the general membership meeting at the same level. When the congress or the general membership meeting is not in session, it is the committee elected by the congress of the general membership meeting.

Article 12. When necessary, the Central Committee may call representative meetings to discuss and decide on major questions that should be solved promptly. The number of representatives at such meetings and the method for selecting the representatives should be determined by the Central Committee or its Standing Committee.

When necessary, the committees of provinces, autonomous regions, and municipalities may call representative meetings to discuss and decide on major questions that should be solved promptly. The number of representatives to such meetings and the method for selecting the representatives should be determined by the committees calling the representative meetings.

Chapter III: Central Organizations

Article 13. The National Congress of the party is held once every five years and convened by the Central Committee. It may be held before the due date or postponed if the Central Committee deems it necessary to do so. The number of delegates to the National Congress and the procedure governing their election should be determined by the Central Committee or its Standing Committee.

Article 14. The functions and powers of the National Congress are as follows:

1. To examine and discuss the reports of the Central Committee;
2. To discuss and decide on the principles, tasks, and other major questions of the party;
3. To revise the constitution of the party;
4. To elect the Central Committee.

Article 15. The Central Committee is elected for a term of five years. However, when the next National Congress is convened before or after its due date, the term should be shortened or extended correspondingly. The number of members and alternate members of the Central Committee should be determined by the National Congress.

When the National Congress is not in session, the Central Committee carries out the decisions of the National Congress, directs the work of the whole party, and represents the party in its external relations.

The Central Committee meets in plenary sessions once a year, and such sessions are convened by its Standing Committee. If necessary, they may be held before or after the due date.

Article 16. The functions and powers of the plenary meeting of the Central Committee are as follows:

1. To examine and approve reports of the Standing Committee of the Central Committee;
2. To discuss and decide on important matters concerning the party;
3. To elect the chairman, vice-chairmen and standing committee members of the Central Committee;
4. To elect the chairman, vice-chairmen and members of the Executive Bureau of the Central Committee;
5. To choose the secretary general, directors of departments, and chairmen of various committees of the Central Committee;
6. To choose advisers to the Standing Committee of the Central Committee and advisers to the Central Committee.

Article 17. The Standing Committee of the Central Committee is composed of the chairman and vice-chairmen of the Central Committee and members of the Standing Committee and directs the work of the whole party when the plenary meeting of the Central Committee is not in session.

The Standing Committee of the Central Committee meets once every three months, and such meetings are convened by the chairman of the Central Committee and may be held before or after the due date if necessary.

Article 18. The Executive Bureau of the Central Committee is composed of a chairman, vice-chairmen, and members. It presides over the day-to-day leading work of the Central Committee under the leadership of the Standing Committee of the Central Committee.

Article 19. Under the leadership of the Executive Bureau of the Central Committee, the secretary general of the Central Committee presides over the day-to-day work of the central organs, organizes the implementation of the resolutions and decisions of the Executive Bureau of the Central Committee, and links, coordinates, and expedites the work of various departments of the Central Committee.

The Central Committee has a number of deputy secretaries general to assist the secretary general in his work. The choice of deputy secretaries general is decided by the Standing Committee of the Central Committee.

Article 20. The Central Committee establishes the necessary working departments. The choice of deputy directors and vice-chairmen of the various departments is decided by the Standing Committee of the Central Committee.

Article 21. Both the Standing Committee of the Central Committee and the Central Committee may have advisers.

Article 22. The Central Committee establishes a Unity Committee. The Unity Committee has a chairman, vice-chairman, and members. The choice of vice-chairmen and members of the Unity Committee is decided by the Standing Committee of the Central Committee.

Chapter IV: Local Organizations

Article 23. Party congresses or general membership meetings of provinces, autonomous regions, municipalities directly under the central government, or cities divided into districts are held once every five years. A party congress or general membership meeting of a city (county and municipal district) is held once every three years. All are convened by the party committees at the corresponding levels. When necessary, they may be held before or after their due dates upon approval by the next higher party organizations.

The number of delegates to the local party congresses and the procedure governing their election are determined by the party committees at the corresponding levels.

The functions and powers of local party congresses at various levels are as follows:

1. To examine and approve reports of party committees at the corresponding levels.

2. To discuss and decide on important matters concerning the party committees at the corresponding levels;

3. To elect the party committees at the corresponding levels.

Article 24. Local committees at various levels shall each elect a chairman and several vice-chairmen.

The provincial, autonomous region, and municipal committees and the committees of cities with districts under them may, according to the needs of their work, each elect a Standing Committee consisting of a number of Standing Committee members and a secretary general.

The results of the elections of chairman, vice-chairmen, Standing Committee members, and secretary general should be reported to higher-level organizations for the record.

Article 25. Local committees at various levels may set up working departments according to the needs of their work.

The provincial, autonomous region, and municipal committees and the committees of cities with districts under them may have advisers, if necessary. The choice of advisers is decided by the committees at the corresponding levels.

Article 26. The regulations governing the organization of local committees at various levels shall be formulated by the Central Committee separately.

Chapter V: Primary Organizations

Article 27. This party's primary organizations are the chapters. A basic unit where there are five or more party members may form a chapter with the approval of the higher level organizations. A basic unit where there are three or four party members may form a cell, or have the party members join another chapter in a nearby area, or join a chapter of a similar profession [i.e., members may join a chapter whose members are engaged in work comparable to the member's profession].

Article 28. The committee of a chapter is elected by a general membership meeting every other year. If necessary, the election may be held earlier or later with the approval of higher level organizations. The number of committee members is decided by the next higher level organization.

A chapter committee shall elect a chairman and, if necessary, vice-chairmen.

Article 29. The main tasks of the party organizations of this party are:

1. To propagate and implement the principles and policies of the Communist Party of China and the government; to convey to the party members and implement the resolutions of higher level organizations and the tasks assigned by them; to maintain close contacts with the masses, report on their situation, and put forward opinions and make suggestions to higher-level organizations;

2. To bring into full play the initiative and creativeness of the party members and the persons with whom the primary organizations maintain contacts, encourage them to make more contributions to socialist modernizations and to the reunification of the motherland; and to discover talented persons, commend advanced persons, and popularize advanced experience;

3. To urge and organize the party members to study and constantly raise their ideological and political understanding and their ability to serve the socialist cause;

4. To enforce party discipline and conduct criticism and self-criticism;

5. To reflect the party members' opinions and demands;

6. To recruit new party members, collect membership dues, and discuss rewards for or punishment of party members.

Chapter VI: Supplementary Article:

Article 30. The power to revise this constitution belongs to the National Congress; the power to interpret it rests with the Central Committee.

Notes

Full citations for most items in the notes may be found in the bibliography. The following abbreviations are used in the notes and bibliography:

BJRB	*Beijing ribao* (Beijing Daily)
CB	*Current Background*
CNA	*China News Analysis*
DGB	*Da gong bao* (Workers Daily)
ECCM	*Extracts from China Mainland Magazines*
FBIS	Foreign Broadcast Information Service
FRUS	*Foreign Relations of the United States*
GM	*Guangming ribao* (Guangming Daily)
GR	*Gongren ribao* (Daily Worker), Beijing
JFRB	*Jiefang ribao*
JPRS	Joint Publications Research Service
PD	*Renmin ribao* (People's Daily)
SCMP	*Survey of China Mainland Press*
SW	*Selected Works of Mao Tse-tung*
WD	*Workers Daily* (Gongren ribao)
WHB	*Wen hui bao*
Xinhua	Xinhua News Agency
ZXS	*Zhongguo xinwen she*

Chapter 1

1. George Konrád and Ivan Szeléyi, *Intellectuals on the Road to Class Power* (New York: Harcourt, Brace, Jovanovich, 1979).
2. In the 1950s, Chinese theorists tried to find sanction for a multiclass socialist state in the writings of Lenin, but the results do not strike one as very satisfactory. See Hu Xiyuan,"Why Do We Adopt the Policy of Long-term Coexistence?"; Liu Shaoqi, "Political Report"; Zhang Zhiyi, "On the Recognition of the

Policy of Long-term Coexistence"; and You Guangyuan, "The Class Nature of the Present People's Democratic Dictatorship." The latter cites Lenin, "The Proletarian Revolution and the Renegade Kautsky," *Collected Works*, vol. 23 (New York, 1945).

3. Leonard Schapiro, *The Origin of the Communist Autocracy: Political Opposition in the Soviet State: First Phase, 1917-1922* (Cambridge, Mass., 1955), 206.

4. See Steven Kertesz, ed., *The Fate of East Central Europe*, 259, 287, 299.

5. See Oscar Halecki, "Poland," in ibid., 134 f.

6. J. P. Nettle, *The Eastern Zone and Soviet Policy in Germany, 1945-1950* (London, 1951); Kurt Glaser, "Governments of Soviet Germany," in *Government in Postwar Germany*, ed. Edward H. Litchfield et al. (Ithaca, 1953); Thalheim, "East Germany," in Kertesz, *Fate of East Central Europe*.

7. Edward Taborsky, "Non-Communist 'Parties' in Czechoslovakia," *Problems of Communism* (March-April 1959), 20.

8. For an official interpretation, see Li Weihan, "The United Front Work and the Party," Xinhua News Agency, September 25, 1956, *CB* 418.

9. Mao Zedong, *Selected Works* (Beijing, 1964), 1:13-21.

10. Ibid., 2:321.

11. This term sometimes includes the peasantry, but it will not be used that way in this book.

12. Mao, *SW*, 2:322.

13. Mao, *SW*, 3:339-384.

14. Mao, *SW*, 3:34.

15. Gunther Stein, *The Challenge of Red China*, 466.

16. Mao, *SW*, 3:255-320.

17. Andrei Zhdanov, "The International Situation," *For a Lasting Peace, for a People's Democracy!*, November 10, 1947.

18. Quoted in John H. Kautsky, *Moscow and the Communist Party of India: A Study in the Post-War Evolution of International Communist Strategy* (New York, 1956), ch. 2.

19. Ibid. The issues were much more complicated than I indicate here. A "leftist" program was really a "united front from below"--toward which a few members of the bourgeoisie might be attracted.

20. Liu Shaoqi, "Internationalism and Nationalism." Whereas the Soviet view is perhaps academic as far as this study of China is concerned, it is not so in the case of Communist

movements elsewhere. In India, the failure of leaders to interpret correctly the zigzags in the Moscow line caused a series of crises within the Communist Party and brought the fall of Ranadive from his position as Party head. See Kautsky, *Moscow and the Communist Party of India.*
21. See Benjamin I. Schwartz, "Ideology and the Sino-Soviet Alliance," in Howard L. Boorman, ed., *Moscow-Peking Axis: Strengths and Strains* (New York, 1957), 129-31.

Chapter 2

1. *CB* 327, p. 1; Melville T. Kennedy, "The Chinese Democratic League," 144.
2. See *Zhongguo qingnian dang shih lue ji zhengkang*; and Liu Dongyan, "My View of the China Youth Party's Past, Present, and Future."
3. Also known as the China Revolutionary Party, the Provisional Action Committee of the Guomindang, and the Chinese National Liberation Action Committee. For more information on the Third Party, see Wang Jingwei et al.
4. In 1951 the Communists executed Li Xiyuan for his alleged partial responsibility in the incident.
5. Albert H. O'Bryant, "Liang Sou-ming: His Response to the West," *Papers on China from the [Harvard] Regional Studies Seminars,* vol. 7 (Cambridge, Mass., 1953), 3.
6. See Carsun Chang, *Third Force in China,* 25; Carsun Chang, *The Democratic Socialist Party's Platform Explained*; Kennedy, "Chinese Democratic League," 143; and Lu Yi'an, *The Split in the Democratic Socialist Party.*
7. Usually translated "Seven Gentlemen"--an inappropriate rendering in view of the fact that one (Shi Liang) was a woman.
8. Chang, *Third Force,* 80; Kennedy, "Chinese Democratic League," 143f.
9. Hong Kong was occupied by the Japanese from December 1941 through the summer of mid-1945.
10. Some in the Nationalist Party apparently believed that the League was appealing to labor union members, but there is no evidence of this. Certainly there was virtually no peasant representation.
11. During the 1940s it was often unclear whether the League was a federation of parties, or whether ultimately it comprised

members of those parties and other individuals. At first it was too heterogeneous for the latter to be the case, but after the Youth Party and National Socialists (Democratic Socialists) left, it became more a party of individuals without intermediate organs.

12. The information in this paragraph comes from a short biography of Zhang Lan in Xiao Yena and Wang Erde, *Communist China's Democratic Parties*, 34-36.

13. See Roderick MacFarquhar, *Origins of the Cultural Revolution*, 1:277.

14. Peter S. H. Tang, *Communist China Today: Domestic and Foreign Policies* (New York, 1957), 150.

15. Stein, *Challenge of Red China*, 464-66. I have modernized the spelling.

16. The resulting CEC had about sixty members, half from the six constituent groups and the remainder representing various professions, etc.

17. Chinese Nationalist sources describe Zhou as an "undercover Communist." *Issues & Studies* (May 1984), 76.

18. On the Democratic Socialist Party, see Chang, *Third Force*, Howard L. Boorman, *Biographical Dictionary of Republican China*, 1:33-4, and works cited above in note 6. There is no such account in English of the China Youth Party, which is still actively supporting the Guomindang. For example, in 1986 CYP Chairman Li Huang condemned foreign support for Taiwan's democracy movement. *China Post* (Taipei), May 22, 1986.

19. This is not to say that the Guomindang totally lost interest in winning over the group. A liaison, Lei Zhen, was appointed to negotiate with Luo Longji, a leading figure in the League. (Lei eventually went to Taiwan where he tried to establish a "China Democratic Party." He was then imprisoned for ten years.)

20. Including General George Marshall (see below) and Ambassador John Leighton Stuart. (See his *Fifty Years in China*.) In 1946, the two inspired President Truman to send a confidential message to Chiang Kai-shek criticizing the "cruel murders" of Chinese liberals and the government's increasing tendency to "resort to force, military or secret police." Today called "quiet diplomacy," the démarche failed.

For their part, the Communists appear not to have objected to the democratic parties receiving help from the Western democracies. Later, Liu Shaoqi would be accused of having advocated that the United States and Britain help the Democratic League. See "Capitalism of China's Khrushchev," Xinhua, July 8, 1967, *SCMP* 3978, 17.

21. Chinese government statement of October 28, 1947. Translated in *United States Relations with China* (Washington, D.C., 1949), 839.

22. The documents from this third plenary session of the CEC are contained in the pamphlet *Zhongguo minzhu tongmeng san zhong quan hui.* See also Allen B. Cole, "The United Front in the New China," 27f.

23. On the RG's roots in the 1920s, see Eugene Z. Hanrahan, "The Birth of the Chinese Red Army," ms., 1953, in Columbia University Libraries, 107. In August 1957, after Luo Longji had fallen out of favor with the CP, the Xinhua News Agency asserted that in 1948 in Shanghai he was supposed to send money to DL headquarters in Hong Kong but failed to do so. Apparently the League was having financial difficulties at this time. Xinhua, August 10, 1957, *CB* 475. For more information on the League before 1949, see *Heping minzhu tongyi jianguo zhi lu*; *Minzhu tongmeng wenxian*; *Minzhu tongmeng er zhong quan hui zhengzhi baogao*; and *Forty Years of the China Democratic League.*

24. Xiao and Wang, *Communist China's Democratic Parties*, give 1942 as the date of the founding of the two associations, but other sources agree on 1946. Li Jishen, "My Party and What It Stands For"; Cole, "United Front," 48.

25. *Asiaweek*, July 12, 1985, 26.

26. See Harold R. Isaacs, *The Tragedy of the Chinese Revolution* (Stanford, 1961), ch. 17.

27. *Foreign Relations of the United States* (Washington, D.C., 1948), 7:305, 313, 337, 354, 364f.

28. Xiao and Wang, *Communist China's Democratic Parties*, 11. Li's name is sometimes rendered according to the Cantonese reading of the characters: Li Chai-sum.

29. Ibid., 51.

30. Ibid., 63. On the early Jiusan, see also *Gongren ribao*, Beijing, September 12, 1956, *SCMP* 1414; Xinhua, February 11, 1956, *SCMP*, 1230; Xinhua, June 14, 1956, *SCMP* 1312; and Xi Zhongxun, "Speech Commemorating the Fortieth Anniversary of the Jiusan."

31. Xiao and Wang, *Communist China's Democratic Parties*, 54.

32. See *United States Relations with China*, 201f., 675f.

33. Apparently the new name was first suggested by Zhang Bojun in 1945. See Yan Xinming, "Listen to Zhang Bojun's Delirious Statement." (Not everyone approved of this meaningless new name, and it was not until 1947 that it was formally adopted.)

34. *United States Relations with China*, 214.

35. For general information about the middle parties before 1949, see Yang Hanhui, *Political Education in Contemporary China*; Zhang Zhiyi, *Political Parties and Groups*; *Zhongguo xin minzhu yundong zhong de dangpai*; *Zhongguo ge xiao dangpai xiankuang*; Yu Runtang and Yao Chuankeng, *Contemporary Political Parties in China*; and Ping Xin, *On the Third Force and the Democratic Movement*.

36. I have said little about the programs of these parties. They (or at least the League) were mainly distinguished from the two major parties by a more sincere dedication to democracy--though all parties professed democratic intentions. In terms of economic program, all of China's parties favored socialism to one degree or another, but none had a coherent program except the Communists, who radically changed their policies as soon as they gained power. For his part, Zhang Lan did not consider it appropriate to make controversial philosophical or programmatic statements; rather, his main concern was to prevent the Democratic League from fragmenting.

37. Stein, *Challenge of Red China*, 6

38. Cole, "United Front," 39.

39. Ibid., 41.

40. In 1948, Pan Zugang had been removed as head of the Communist headquarters in Hong Kong because of his hostility to DL and RG figures. *FRUS*, 1948, 7:270.

41. Cole, "United Front," 29.

Chapter 3

1. Donald W. Klein and Anne B. Clark, *Biographic Dictionary of Chinese Communism*, 1:534-40. Li had also gone by the name Lo Mai.

2. Only a particularly large chapter is divided (into cells). The term "branch" is being avoided because in English it sounds as though it includes multiple chapters. To convey this latter sense (e.g., to indicate all the chapters of a municipality),

I will use the word "section."

3. Because hardly any DPG members have been women (see chapter 7, note 25), masculine pronouns will be used in this book.

4. See appendix 3 for the text of the RG constitution, as revised in 1983.

5. On DPG finances, see chapter 7.

6. Li Junlong had studied at Columbia University and then served in the Nationalist government. He later became a member of the RG Standing Committee but was declared a "rightist" during the late 1950s.

7. Xiao and Wang, especially 23-27.

8. Sometimes a member of the People's Liberation Army could join the RG, though normally PLA people have not been eligible for DPG membership.

9. For details, see Lyman P. van Slyke, *Enemies and Friends*, 214.

10. *Renmin ribao* (People's Daily) (*PD*), July 23, 1953, *SCMP* 644. Zhang Bojun, head of the PW, was also a key figure in the Democratic League.

11. Huang Yanpei, *The Past Eighty Years*, 102. Zhou Enlai is said to have made the arrangements (from Beijing).

12. *Da gong bao* (Workers Daily) (*DGB*), Tianjin, April 14, 1955, *CB* 327. The quote is from a later NCA constitution, but recruitment policies have been consistent. Membership in more than one group was common, and it was not unheard of for a person to belong to the CP and a DPG at the same time. See *CB* 327, 2; Xinhua, May 8, 1957, *SCMP* 1543; Xinhua, August 10, 1957, *CBG* 475. In addition, there is evidence that Communists were secretly planted in the DPGs. Xinhua, July 7, 1957, *SCMP* 1572. A member could not quit his DPG without special permission. These questions will be discussed further in chapter 7.

13. Xinhua, February 3, 1951, *SCMP* 62.

14. Xiao and Wang, *Communist China's Democratic Parties*, 9-10, 31-34, 51, 52; *China News Analysis*, 124. The NCA had a special Industrial Reconstruction Guidance Department to assist in the control and reform of members.

15. *PD*, August 19, 1957, *SCMP* 1606.

16. For early examples, see Xinhua, February 29, 1956, *SCMP* 1242; Li Jishen, "My Party." Note that another group serves a related purpose--the Taiwan Democratic Self-Govern-

ment League.

17. On the case of Wei Lihuang, see Xinhua, March 16, 1955, *SCMP* 1009; Xinhua, March 17, 1955, *SCMP* 1010. More recent illustrious defections from the Nationalists have been Li Zongren and Ma Bi.

18. Note that another group, the Zhi Gong Dang, serves to link overseas Chinese to the Beijing government.

19. Liu Shaoqi, "Report on the Draft Constitution," *People's China*, October 1, 1954, 32f.

20. *Extracts from China Mainland Magazines* 87, 9; Li Jishen, "My Party"; Xinhua, March 27, 1956, *SCMP* 1264.

21. See *CB* 327, 12; Xinhua, March 4, 1957, *SCMP* 1484; Faure, 89.

22. *Xin Hunan bao*, May 14, 1955, *SCMP* 1074 Suppl.; Xinhua, January 9. 1957, *SCMP* 1453.

23. Quoted in William Stevenson, *The Yellow Wind: An Excursion In and Around Red China with a Traveler in the Yellow Wind* (Cambridge, Mass., 1959), 134.

24. Xinhua, November 22, 1953, *SCMP* 695.

25. But no deputy premierships in the Supreme State Council. On this, see Chu Anping, "Allow Me to Offer Some Opinions." The NPC-SC and SSC were both post-1954 organs. Before then there was the Central People's Government Council, with several non-Party vice-chairmen.

26. For example, Wang Zhixiang. A Jiusan legal scholar, Wang participated in the drawing up of the constitution, the draft criminal code, and other laws. Xinhua, January 29, 1986, FBIS, Jan. 31, 1986.

27. Li Weihan, "Further Strengthening the United Front Work."

28. On the case of executed RG figure Zhou Yixiang, see Cheng Peng, "An Injustice Hidden for Thirty Years."

29. During the 1950s, the DPGs were among those targeted during the following campaigns: Land Reform (1950-53); Resist-America, Aid Korea (1950-53); Three-anti (1951-52) and Five-anti (1952); Judicial Reform Movement (1952-53); Propaganda Educational Work of the General Line During the Transition Period (1953-54); Movement for the Extermination of All Hidden Counterrevolutionaries (1955-56); Rectification Movement for the Whole People (1957-58); Heart-delivering (*Jiao xin*) Movement (1958).

30. Xinhua, November 29 and 30, 1951, *SCMP* 226.

31. For source citations, see James D. Seymour, "Communist China's Bourgeois-Democratic Parties," 50, n. 20.
32. Xinhua, March 9, 1952, *SCMP* 291.
33. Three-anti was directed against corruption, waste, and bureaucratism; Five-anti concerned bribery, tax evasion, stealing state property, delivering substandard goods to the government, and improperly obtaining official information.
34. For source citations, see Seymour, "Bourgeois-Democratic Parties," 51, n. 23.
35. For source citations, see ibid., n. 24.
36. Xinhua, November 30, 1951, *SCMP* 226; Xinhua, November 29, 1951, *SCMP* 226.
37. Kirby, *Contemporary China* (Hong Kong, 1958), 2:209.
38. Xinhua, January 20, 1953, *SCMP* 497.
39. *Guangming ribao* (*GM*), May 26, 1957, quoted in MacFarquhar, *The Hundred Flowers*, 210.
40. On this problem, as viewed by members of various DPGs, see MacFarquhar, *The Hundred Flowers*, 85, 97, 100, 105, 109f.
41. *Renmin shouci*, 1955, cited in *CB* 327, 38.
42. For source citations, see Seymour, "Communist China's Bourgeois-Democratic Parties," 53, n. 31.
43. For source citations, see ibid.
44. See NCA Hundred Flowers comments in MacFarquhar, *The Hundred Flowers*, 198f., 204, and 206.
45. For source citations, see Seymour, "Communist China's Bourgeois-Democratic Parties," 55, n. 32.
46. Xinhua, January 29, 1956, *CB* 376. Zhou also called for more intellectuals to be admitted into the Communist Party.
47. Xinhua, January 30, 1956, *CB* 325.
48. Li Weihan, "Our People's Democratic United Front."
49. This was quoted in a later article attacking Liu. Shi Pai, "Refuting Several 'Bases of Argument' for the Theory of 'Class Cooperation' of the Number One Ambitionist in the Party," *GR*, June 29, 1967, *SCMP* 3987, p. 2.
50. *GM*, March 2, 1956, *SCMP* 1246.
51. Richard Hughes, recalling an interview with Sa Gongliao in 1956 or 1957. *Far Eastern Economic Review*, January 20, 1978, 25. Sa claimed input in matters of detail. "There is so much detail. You should hear our suggestions when we consult. . . . On policy lines, the Communist Party is making no mistakes at all. None. Let us be clear on that."

52. See the sentiments of Jiusan member Qian Chuchun, quoted in MacFarquhar, *The Hundred Flowers*, 99.

Chapter 4

1. Shi Rugang, "My Understanding of 'Long-term Coexistence.' "
2. Hu Xiyuan, "Why Do We Adopt the Policy of Long-term Coexistence."
3. Liu Shaoqi, "Political Report."
4. Li Weihan, "United Front Work."
5. *GM*, September 15, 1956, *SCMP* 1382.
6. Ibid., and Xinhua, September 1, 1956, *SCMP* 1382.
7. Mao Zedong, *SW*, 4:411.
8. *PD*, July 25, 1956, cited in Hinton, "The 'Democratic Parties.' "
9. Gu Zhizhong, "Relations Between the Communist Party and the Democratic Parties."
10. Li Weihan, "The United Front Work and the Party."
11. Ibid. Elsewhere Li even cites Mao's 1941 "cross fire" statement. Li Weihan, "Our People's Democratic United Front."
12. At the government's request, the CPPCC did carry out inspection trips between the autumn of 1955 and the spring of 1957.
13. The pre-1949 size of the democratic parties is unknown. Stein (p. 372) suggests that the membership of the DL alone may have approached 100,000, but the New China News Agency puts the total 1949 DPG population at 20,000.
14. For source citations, see Seymour, "Communist China's Bourgeois-Democratic Parties," 66f, notes 14-20.
15. For source citations, see ibid., 68, n. 21.
16. Xinhua, August 23, 1957, *SCMP* 1606.
17. *GM*, August 9, 1957, *SCMP* 1600.
18. Xinhua, August 10, 1957, *CB* 475.
19. Luo Longji.
20. Li Weihan, "Democratic Parties."
21. For source citations, see Seymour, "Communist China's Bourgeois-Democratic Parties," 69, n. 27.
22. Xinhua, June 25, 1957, *SCMP* 1571.
23. Xinhua, June 15, 1957, *SCMP* 1558.
24. Zhang Bojun, "I Bow My Head." (Similar remarks of Zhang are quoted in Appendix 2.)

25. Xinhua, December 14, 1957, *SCMP* 1694.
26. Wen Yiwen, "'Pernicious Expansion' Has Corrupted the Jiusan Society," *GM*, August 29, 1957, *SCMP* 1623.
27. Xinhua, August 28, 1957, *SCMP* 1611.
28. Wen Yiwen, "'Pernicious Expansion.'"
29. E.g., Harbin's APD section, established in 1957. No Heilongjiang provincial APD organization was established for almost twenty years. Harbin Radio, October 28, 1983, JPRS 84857, 93.
30. There is an inconsistency here with Zhang's statement quoted above. "Ten thousand" probably simply means "a great many."
31. Xinhua, August 16, 1957, *SCMP* 1600.
32. Xinhua, February 22, 1957, *SCMP* 1485.
33. Xinhua, August 15, 1957, *SCMP* 1606.
34. I am reliably informed that the total DL membership in December 1957 was 33,188. I surmise that this was after a few thousand had been dismissed. See below chapter 5, note 76.
35. For source citations, see Seymour, "Communist China's Bourgeois-Democratic Parties," 72, n. 40.

Chapter 5

1. Zhou Enlai, "On the Question of Intellectuals."
2. Mao Zedong, "On the Ten Great Relationships" (talk delivered in April 1965, but text not widely circulated until the winter of 1965-66), in *Mao*, ed. Jerome Ch'en (Englewood Cliffs, N.J., 1969), 77-78.
3. The reference is to the "hundred schools" during the Warring States period (481-221 B.C.), when there was cultural pluralism and flourescence.
4. In the interim, three DPG figures were made ministry heads: Luo Longji (timber), Xu Dehong (marine products), and Li Zhuchen (food industry). Also, *Guangming Daily* called for freer discussion. Harold C. Hinton, *Contradictions in Communist China*, 5.
5. Lu Dingyi, "Let All Flowers Bloom Together."
6. *CB* 412, 61.
7. The date of Mao's essay "On Contradiction" (*SW*, 1:311-47), officially "August 1937," is in dispute. Some say that it must have been written later than that, but Mao reaffirmed the date in a 1964 interview with Edgar Snow. *New Republic,* Janu-

ary 20, 1965.
  8. Mao, *SW*, 1:315.
  9. Ibid., 1:322.
  10. Mao, *On the Correct Handling of Contradictions,*, 22f.
  11. Ibid., 52.
  12. Ibid., 53.
  13. *On the Correct Handling of Contradictions*, in the version published in the late spring of 1957, contains six criteria by which to determine whether a criticism should be considered valuable to the state. Their common denominator was compatibility with CP leadership. It is generally assumed that these were not in the speech as it was delivered in February. Whether they were or not is not crucial for our purposes since even in the final version Mao does not actually prohibit the expression of views that fail to meet these criteria. The latter are simply offered as a measure for evaluation. Furthermore, even if the six criteria were not in the original speech, it is unlikely that no limits were suggested. In April, Deng Chumin, writing in *Guangming Daily*, outlines four "circles" of his own to which the malcontent were urged to restrict themselves. Deng, a member of the DL, had heard Mao's address. If the latter had contained no suggestions for limitations of criticism, it is unlikely that he would have outlined any. On possible differences between the two versions of Mao's speech on the subject of the universality of contradictions (i.e., their applicability to the Soviet Union), see Hinton, *Contradictions,* 20. For Mao's earlier thinking on the subject, see "On Contradiction" (dated 1937), *SW*, 1:311-47.
  14. Mao, *SW*, 5:408-14.
  15. This seems to be the view taken by G. F. Hudson, who relies on the following item taken from *People's Daily* (July 1, 1957): "From May 8 to June 7 the newspapers of the Chinese Communist Party, following the directive of the Party's Central Committee, published few or no affirmative views or counter criticisms. The Party foresaw that a class battle between the bourgeoisie and the proletariat was inevitable. For a long time, in order to let the bourgeoisie and the bourgeois intellectuals wage this battle, we . . . did not counter the frantic attacks made by the reactionary bourgeoisie rightists. The reason was to enable the masses to distinguish clearly between those whose criticism was well intended and those who were inspired by ill will. In this way the forces for an opportune counterblow

amassed their strength. Some people call this scheming, but we say it was quite open. We told the enemy in advance that before monsters and serpents can be wiped out, they must first be brought into the open, and that only by letting poisonous weeds show themselves above ground can they be uprooted." This, it would seem, does not prove what Hudson says it does. Surely it was not necessary to go to the extremes of May 1957 to spot those dissatisfied with the Party line. *People's Daily* is simply rationalizing after the fact. The "we planned it all along" explanation is not to be taken seriously.

16. The two forums ran from May 8 to 16 and May 21 to June 2. See below, n. 52.

17. Xinhua, May 8. 1957, *SCMP* 1543. It is likely that in the spring of 1957 some top official privately exhorted DPG people to criticize the Party. After all, Mao had first spoken of Hundred Flowers over a year earlier, but until this moment nearly everyone had exercised extreme caution.

18. Chen Qing, "Refutation of Zhang Naiqi's 'My Self-Examination.' "

19. Xinhua, June 3, 1957, *SCMP* 1548 (Xinhua paraphrase).

20. Xinhua, May 19, 1957, *SCMP* 1543.

21. Xinhua, June 8, 1957, *SCMP* 1560.

22. Xinhua, June 6, 1957, *SCMP* 1550.

23. The term used by DL member Yang Qing of Qilin, quoted in MacFarquhar, *The Hundred Flowers*, 110.

24. Xinhua, June 5, 1957, *SCMP* 1548.

25. MacFarquhar, *The Hundred Flowers*, 110.

26. Xinhua, June 1, 1957, *SCMP* 1550.

27. Xinhua, May 8, 1957, *SCMP* 1543.

28. Xinhua, May 11, 1957, *SCMP* 1543; Xinhua, June 20, 1957, *SCMP* 1547.

29. MacFarquhar, *The Hundred Flowers*, 226.

30. Xinhua, May 10, 1957, *SCMP* 1543.

31. See also Xinhua, June 15, 1957, *SCMP* 1558 (Mao likened to Qin Shi Huangdi, and the Party in general to Yuan and Qing dynasty rulers). On remarks relating to the handling of counterrevolutionaries among members of the DPGs, see Xinhua, May 9, 1957, *SCMP* 1543; Xinhua, June 30, 1957, *SCMP* 1571; Xinhua, June 15, 1957, *SCMP* 1558; *PD*, June 19, 1957, *SCMP* 1561; and Luo Longji, "My Preliminary Examination."

32. MacFarquhar, *The Hundred Flowers*, 101f.

33. Xiao and Wang, *Communist China's Democratic Parties*,

37. See also Xinhua, July 3, 1957, *SCMP* 1571.
34. Xinhua, July 3, 1957, *SCMP* 1571.
35. Some people went even further than this, urging rotation of power among the Communist and democratic parties. MacFarquhar, *The Hundred Flowers*, 123. See also p. 227.
36. Boorman, *Biographical Dictionary*, 2:437.
37. MacFarquhar, *The Hundred Flowers*, 170.
38. Xinhua, July 4, 1957, *SCMP* 1569.
39. Zhang Naiqi, "Several Questions Concerning the Work of Assisting and Guiding the Transformation of Industry and Commerce." This particular distinction by Zhang could not have convinced many socialists, but his general argument that intellectuals and businessmen are workers became the official CP line in the 1980s.
40. Xinhua, May 7, 1957, *SCMP* 1529.
41. Zhang Naiqi, "Work Report of the China National Construction Associaton."
42. Chen Qing, "Refutation." (Zhang denied ever having said this.)
43. MacFarquhar, *The Hundred Flowers*, 196.
44. Aside from the *Guangming Daily* there were a number of other publications claiming to speak for the DPGs, among them the Shanghai *Wen hui bao* (on which see Xinhua, June 30, 1957, *SCMP* 1571) and *Zhengming* (Contending)--a monthly sponsored by the DL (see *GM*, April 18, 21, 26, 1957, *SCMP* 1522, 1541, 1541, respectively). For the RG there was *Tuanjie bao* (Unity) (see *SCMP* 1600, p. 14). The RG also had various local publications, usually called *Min-ge*, a Chinese abbreviation for RG. (E.g., Xinhua, July 7, 1957, *SCMP* 1572.) Finally, Construction put out *Min-xin*.
45. But see *PD*, June 11, 1957, *SCMP* 1558.
46. Chu, who was associated with at least two DPGs, was only given this post in April 1957. He sent reporters to various cities to hold forums.
47. *GM*, June 21, 1957, *SCMP*, 1566.
48. Zhang, "I Bow."
49. *GM*, June 21, 1957, *SCMP*, 1566.
50. *PD*, June 2, 1957, *SCMP*, 1550.
51. *Tuanjie bao*, May 15, 1957, *SCMP* 1600.
52. What I have been calling the May Forums ended June 3. Another forum with ostensibly the same purpose was held June 8-19, and it was announced that forums were to be held fort-

nightly thereafter. After June 3, however, the DPGs were no longer aiding the CP in its rectification, but were busy rectifying themselves.

53. *PD*, June 11, 1957, *SCMP* 1558.
54. Xinhua, June 11, 1957, *SCMP* 1553; *GM*, June 12, 1957, *SCMP* 1561.
55. *PD*, June 13, 1957, *SCMP* 1560.
56. *Da gong bao*, June 9, 1957.
57. *GM*, June 10, 1957, *SCMP* 1563.
58. Xinhua, June 18, 1957, *SCMP* 1558. He was not dismissed as minister of food until January 1958. Xinhua, January 31, 1958, *SCMP* 1760.
59. *PD*, June 14, 1957, *SCMP* 1558.
60. Various Xinhua dispatches in *SCMP* 1558.
61. *PD*, June 19, 1957, *SCMP* 1561.
62. Xinhua, June 18, 1957, *SCMP* 1572; *PD*, June 19, 1957, *SCMP* 1561. Long Yun also maintained that China should give less financial aid to other countries. Land reform, said Long, had led to rural bankruptcy. Many good people had been wronged during the suppression of the counterrevolutionaries. Also, there was much waste in construction efforts, and unemployment was serious. Long criticized his colleagues for being soft in their comments regarding the CP. Xinhua, May 25, 1957, and *PD*, May 23, 1957--both in *SCMP* 1548.
63. For example, Chu Yuxian (APD) criticized the mechanical copying of Soviet experiences, especially in education. MacFarquhar, *The Hundred Flowers*, 91.
64. Xinhua, June 21, 1957 *SCMP* 1564.
65. Xinhua, June 23, 1957, *SCMP* 1569.
66. Xinhua, July 6, 1957, *SCMP* 1572. On local developments in other DPGs at this time, see Xinhua, June 30, 1957, *SCMP* 1569; Xinhua, July 4, 1957, *SCMP* 1569.
67. Xu Daohe, "'Mutual Supervision.'"
68. *GM*, August 9, 1957, *SCMP* 1600.
69. Xinhua, August 15, 1957, *SCMP* 1606.
70. Xinhua, August 16, 1957, *SCMP* 1606. See also Xinhua, August 11, 1957, *SCMP* 1611. For local developments, see Xinhua, August 14, 1957, *SCMP* 1606; Xinhua, August 24, 1957, *SCMP* 1611.
71. This was confirmed by one of my interview subjects, winter 1984-85. This source claimed that the accommodation faction was in the majority; this was not always the case.

72. *GM*, October 21, 1957, *SCMP* 1663. (The "forty-two" figure refers to people later labeled "rightists.")

73. MacFarquhar (*Origins*, 1:277) notes how well-informed Hu was about the internal affairs of another DPG and infers that Hu was fed this information so that he could embarrass Mao Zedong.

74. *PD*, August 29, 1957, *SCMP* 1606.

75. *Da gong bao*, September 12, 1957, *SCMP* 1689.

76. This was said to be only 6 percent of the total membership. These figures (from *GM*, October 21, 1957, *SCMP* 1663) lead us to a plausible total DL membership of 36,700, but see discussion above, chapter 4, note 38.

77. Data from private sources. More than a thousand out of the 6,600 were "counterrevolutionaries" or other types of "reactionaries."

78. Various Xinhua dispatches in *SCMP* 1621.

79. *GM*, October 11 and 21, 1957, *SCMP* 1663.

80. Li Weihan, "Democratic Parties."

81. Xinhua, November 29, 1957, *SCMP* 1694.

82. Xinhua, December 24, 1957, *SCMP* 1694.

83. Xinhua, January 19, 1958, *SCMP* 1699.

84. Xinhua, January 31, 1958, *SCMP* 1706.

85. Ibid.

86. *People's Daily*, January 7, 1957, *SCMP* 1507, 17-19. The intra-CP dispute is discussed in MacFarquhar, *Origins*, especially 1:178ff.

87. MacFarquhar (*Origins*, 1:114ff., 191f.) makes more of this point than I am inclined to. In 1956, Liu said: "We should be adept in benefiting from supervision and criticism by members of the various democratic parties and by democrats without party affiliation." Political Report of the Central Committee, in *Communist China, 1955-1959: Policy Documents with Analysis* (Cambridge, Mass., 1962), 188.

88. The official press reported that Luo had "pointed out" (*zhichu*) that Mao had called for the establishment of rehabilitation committees. See MacFarquhar, *Origins*, 1:273.

89. On the DPGs as a surrogate target for Mao, see ibid., 270-78.

90. Li Xifang, "What Kind of Black Thread Runs Through Wu Han's 'Academic' Activities" (in Chinese), *PD*, May 6, 1966, 3.

Chapter 6

1. More information about the DPGs during the 1960s (and also the 1970s) is contained in a paper by Alan Pauw, "Chinese Democratic Parties as a Mass Organization," from which I have benefited in writing this chapter.
2. Wu was quietly admitted to the CP in March 1957, after having sought admission for many years.
3. See Merle Goldman, "The Unique 'Blooming and Contending' of 1961-62."
4. See Huang Yanpei, *The Past Eighty Years*.
5. Pauw, "Chinese Democratic Parties," 16, citing *Communist China*, volume for 1965, 40.
6. Associated Press, August 26, 1966, cited in *Issues & Studies* (April 1984), 97.
7. *Baltimore Sun* dispatch by Peter Kumpa, November 23, 1966.
8. Most of the information on Wu Han here is from James D. Seymour, "The Policies of the Chinese Communists Toward China's Professionals and Intellectuals" (doctoral diss., Columbia University, 1968), 209-223. Portions relating to Wu's persecution are based on a study by Anne F. Thurston. For more information on him, see Tom Fisher, "Wu Han," and Hu Yuzhi and Li Wenyi, "Cherishing the Memory of Comrade Wu Han."
9. Shi Shaopin, "Criticizing Wu Han's Thrown Spears" (in Chinese), *Hongqi* 6 (1966), 24, citing *Zhongguo qingnian* 2 (February 1950).
10. Yao Wenyuan, "On the New Historical Play 'Dismissal of Hai Rui.' "
11. *Hai Rui ba guan* was defended, for example, by Li Zhenyu in *Beijing ribao*, December 9, 1965, *SCMP* 3669.
12. Wu Han, "Self-Criticism on 'Dismissal of Hai Rui.' "
13. On how historians handled themselves in this situation, see *Wen hui bao*, January 7, 1966. Reprinted in *PD*, January 13, 1966, *CB* 783.
14. *Asiaweek*, July 12, 1985, p. 27. For a general analysis of Zhou's role during the Cultural Revolution, see David Bonavia, "New Myth for a Mandarin: Zhou Enlai's Achievements are Revised and Rewritten," *Far Eastern Economic Review*, January 23, 1986, 30f.
15. Speech delivered at a "report meeting," October 24, 1966; *Long Live Mao Zedong Thought*, translated in *Current Back-*

ground 891, October 8, 1969, 72.

16. *New York Times*, February 28, 1978, 5.

17. See Xinhua, January 18, 1980, FBIS, January 21, 1980, L-1.

18. This assessment is based on only one known case--the offspring of a provincial DPG official. Information drawn from an interview by Richard Madsen.

19. In Zhejiang. Pauw, "Chinese Democratic Parties," 16.

Chapter 7

1. A version of this chapter appeared in *Asian Survey* (September 1986).

2. Pauw, "Chinese Democratic Parties," 17, quoting FBIS, December 29, 1977.

3. For a 1985 statistical survey in one province (Sichuan), see Ma Ding, "Our Province Has Implemented United Front Policy."

4. On the case of Zhou Yixiang, see Cheng Peng, "Injustice." (Zhou was an active underground RG member who disappeared in 1952. His execution was announced in 1955.) Another such case is that of Mai Gongbin, a foundering member of the Revolutionary Guomindang. See: Xinhua, January 18, 1980, FBIS, January 21, 1980, L-1.

5. *Zhongguo xinwen she* (*ZXS*), March 5, 1985, FBIS, March 7, 1895, K-17.

6. December 10, 1983, 3, JPRS CPS-84-003, 4. Two comments are interesting: "In the past, campaigns would start inside the Party [as in this case], and then expand to cover everybody." "Even as I climbed on the train to come to the city (Beijing), my son once again urged me not to say too much. 'Don't forget Grandpa's lesson,' he said. My father was 'capped' as a 'rightist' for speaking out in 1957."

7. There were a few instances of involvement of the DPGs in the campaign against spiritual pollution. E.g., account on Lanzhou Radio, November 4, 1983, FBIS, November 9, 1983, T-1. But in general, the campaign appears to have had little impact on these groups. For example, Guangdong CP secretary reassured a non-Party forum: "Different academic views and disputes between different schools, though incorrect, must not be regarded as spiritual pollution; otherwise, people dare not conduct academic debates." Canton Radio, November 9, 1983,

FBIS, November 15, 1983, P-2.

8. Xinhua, November 26, 1983, FBIS, November 28, 1983, K-17. An example of the same point made locally: Jinan Radio, January 18, 1984, FBIS, January 20, 1984, O-3.

9. Pauw, "Chinese Democratic Parties," 19, citing JPRS 74992.

10. Although the subject lies largely outside the scope of this book, in recent years Taiwan relations has been an important area of concern for these groups. The DPGs mainly involved are the Revolutionary Guomindang (linking up with Chinese Nationalists) and the Taiwan Self-Government League (which aims its pitch at the native Taiwanese). On the latter, see forthcoming work by Bruce Jacobs. For list of the organization's officers and advisers, see Xinhua, December 6, 1983, FBIS, December 9, 1983, U-1. See Xinhua, December 6, 1983, FBIS, December 7, 1983, U-1, for further information on them. For the RG position, see Xinhua, December 21, 1983, FBIS, December 23, 1983, U-2.

11. One vehicle for such plaudits were awards for good work. In the five-year period beginning in 1979, awards were given to nearly 20,000 people, about one-sixth of the total 1984 DPG membership. *China Reconstructs* (April 1984), 30.

12. Zhou Aigui, p. K-1. For an example of provincial-level publicity of a similar nature, see Shenyang Radio, October 24, 1985, FBIS, October 28. 1985. Another form of pro-DPG publicity commonly seen during these years were glowing obituaries and eulogies. Examples: APD Central Committeeman Wu Wenzao (Xinhua, October 1, 1985, FBIS, October 3, 1985); recent Guominding defector from Taiwan Ma Bi (Xinhua, October 15, 1985, FBIS, October 18, 1985); and fellow RG figure Zhang Zhizhong (Xinhua, October 27, 1985, FBIS, October 30, 1985).

13. See Xiao Chong, "When Will the Label 'Peasant Party' Be Removed?--A Look at the 'Intellectualization' of the CP Cadres," *Zhengming* (November 1985), 00.17-19, JPRS, CPS-85-122, pp. 101-106.

14. Yu Gang and Yu Jianze, "A Brief Discussion on the Historical Path of Chinese Democratic Parties."

15. One of the more careful historical-theoretical treatments was that by Chen Ziyun, "On Multiparty Cooperation under the Leadership of the Communist Party." See also "Sincerely Cooperate with Non-Party Personages," *People's Daily* edito-

rial, December 10, 1985, 5, JPRS CPS-86-034, 13-15.

16. Liu Xiaoping and Xu Shuang, "Characteristics and Merits Unique to China's Political System," 14.

17. *CD*, December 10, 1983, 3, JPRS CPS-84-003, 4.

18. Her birth is variously put at 1900 and 1907.

19. Xinhua, September 6, 1985, JPRS CPS-85-102, 49. Women have always been poorly represented among the DPGs. For her part, Shi was an outspoken feminist. In 1957 she complained: "Right up to the present day, people in all walks of life neither think much of nor respect women. Even in the minds of some Communist Party members, the remnants of such feudal thoughts still longer. There are departments which are loathe to take on female comrades; others which pick on the female comrades as the first victims of redundancy." MacFarquhar, *The Hundred Flowers*, 229f.

20. Xinhua, January 31, 1986, JPRS CPS-86-023, 47.

21. Xinhua, September 27, 1985, FBIS, October 1, 1985.

22. Xinhua, November 22, 1983, FBIS, November 23, 1983, K-8. Zhou was the brother of Lu Xun, China's most respected twentieth-century writer. Hu Juewen was reelected chairman of the Central Committee of the National Construction Association. Xinhua, November 21, 1983, FBIS, November 21, 1983. Ji Fang was elected head of the Peasants' and Workers' Party, and Xu Deheng continued on as chairman of the Central Committee of the Jiusan. *China Reconstructs* (April 1984), 31.

23. Xinhua, July 2, 1984, JPRS CPS-84-050. The first course of instruction lasted only ten months.

24. Inasmuch as the DPGs are largely urban organizations, it is not surprising that there are few ethnic minority people in them. Still, the subjects insisted that there was no discrimination, and indeed, minorities may not be underrepresented at all. A number of these chapters have a Hui member, and others have Mongols and Manchus. With a single non-Han, a chapter would probably reflect the overall population in this respect.

25. DPG leaders have included a few illustrious women, including Song Qingling (Soong Chingling, of the RG), Shi Liang (see above, note 19), He Xiangning (widow of Liao Zhongkai and mother of Liao Chengzhi), who headed the RG during the 1960s, and Xie Xuehong, founder of the Taiwan Democratic Autonomous League. But women have been scarce among the rank and file. Only the tiny Jiusan has a substantial percentage of women--about 30 percent. I was told that 20

percent of the Democratic League are women, but my chapters ranged from zero to 15 percent female. The NCA has a special committee on women, but mainly it organizes the wives of businessmen. The APD probably has more women, but in my interviews information was not gathered on this DPG.

26. Two DPGs I did not examine firsthand were the Association for Promoting Democracy (primarily school teachers) and the tiny Peasants' and Workers' Party (health workers, etc.). There are two other groups usually classed together with the DPGs, the Zhi Gong Dang (for overseas Chinese) and the Taiwan Autonomous League. For purposes of this book these latter are not considered DPGs.

27. Virtually all RG members are either former Nationalist Party people or relatives thereof. They need not be senior intellectuals. Faced with extinction if it were to confine its membership to former Chinese Nationalists, the RG has been trying to guarantee a future for itself by enrolling descendants of Guomindang figures. The youngest DPG member to come to my attention was the twenty-two-year-old granddaughter of Ma Hungbin. (He had been a Hui military and political figure in the northwest before 1949). Organization Department head Shao Hengqiu appears unenthusiastic about relying exclusively on Guomindang progeny to fill the ranks of the RG. "Some people say we should look to young people whose parents or grandparents had participated in the Guomindang, but this is a terribly feudal concept. We must come up with a better way to revitalize." *Asiaweek*, July 12, 1985, 27. For more information on the RG, see *Issues & Studies* (April 1984), 92-98.

28. A Taiwan source asserts that the FIC "is actually part of the NCA." *Inside China Mainland* (March 1984), 13.

29. Hangzhou Radio, March 21, 1985, FBIS, March 25, 1985.

30. Shao Hengqiu, quoted in *Asiaweek*, July 12, 1985, p. 26.

31. The rules for disseminating internal information to DPG people were spelled out in both the CPPCC newspaper *Renmin zhengxie bao* and *PD* on March 21, 1984, FBIS, March 26, 1984, K-16 to K-19.

32. Even such "welcome pressure" plays a role in the political process. An article in *Beijing Review* gives these examples of political initiatives from the DPGs: reforming traditional Chinese medicine, restoring classic cuisine, correcting problems in the tea and publishing industries, and improving teacher training programs. See Li Rongxia, "Democratic Parties Work

for Modern China," 21.

33. In 1985 the CPPCC had 2,645 members and a total membership of PPCCs of 280,000. *China Daily*, January 1, 1985.

34. E.g., culture, education, sanitation, economic construction, agriculture, religion, nationalities, athletics, international problems, and women's issues.

35. *China Daily*, April 10, 1985, 4, JPRS CPS-85-041, 44 (article by sociologist Fei Xiaotong about the CPPCC). On (C)PPCC political input, see Xinhua, April 9, 1985, JPRS CPS-85-042, 7; Xinhua April 2, 1985, CPS-85-039, 24; *China Daily*, January 21, 1985, "CPPCC Proposals Play Role"; and "Closer Contacts Between Government Agencies, CPPCC Urged," *Jiefang ribao*, January 26, 1986, 2, JPRS CPS-86-035, 51.

Despite the obvious advantages, some who could be CPPCC members have declined. A Hong Kong man is reported to have rebuffed the offer, saying that he was not pleased by the invitation because the Party only puts its enemies in this body.

36. Quoting a Hubei CP deputy secretary. Wuhan Radio, March 21, 1985, JPRS CPS-85-035, 119. See also Changsha Radio, January 25, 1985, FBIS, January 28, 1985, P-3. Letters may also be sent to officials, which may be answered. In one case, a supportive reply from a provincial first Party secretary was published in the press. E.g., letter from Guo Feng to Zhou Pinwei, *Liaoning ribao,* April 14, 1984, JPRS CPS-84-040, 37.

37. Radio Nanning, June 23, 1984, JPRS CPS-84-050, 12. For other DPG complaints about China's education system, see Xinhua, October 21, 1984, FBIS, October 23, 1984, K-6 to K-8; Xinhua, March 27, 1985, JPRS CPS-85-039, 25f.; and Xinhua, May 5, 1985, FBIS, May 14, 1985, K-17.

38. *Fujian Daily*, January 20, 1984, JPRS, February 2, 1984, O-1.

39. See Kenneth Lieberthal and Michel Oksenberg, "Waiting for the Three Gorges Dam," *The China Business Review* (September-October 1986), 7-9.

40. In the Taiwan Democratic Self-Government League (normally considered a DPG), people may belong to the national organization without belonging to a chapter. However, this rarely appears to be the case with most DPGs. See Chapter I, Article 2 of the Taiwan League's Constitution, Xinhua, December 9, 1983, JPRS 85017, 79.

41. The same was true of rank-and-file people. Journalist Richard Hughes was told by a DL leader in 1956-57: "Some of

our Democratic League members have joined the Party; some Communist members have joined the Democratic League. We work together, you understand." *Far Eastern Economic Review*, January 20, 1978, 25.

42. A Chinese Nationalist source seems to quote the CPPCC journal *Renmin zhengxie bao* (March 19, 1985) as saying that inasmuch as many DPG members are qualified to join the CP, they should be invited to do so. After they have entered the Party, they should continue their DPG activities. A poor translation appears in *Inside Mainland China* (Taipei), November 1985, 15. Other Chinese Nationalist sources (c. 1961) claimed that as many as 50 percent of DPG members had been Communists (with an official minimum of 2 percent), but it is unlikely that the actual figure was ever more than 5 percent. See van Slyke, *Enemies and Friends*, 218.

43. Xinhua, December 13, 1983, JPRS CPS-84-003, 2.

44. In 1957 a Jiusan member reported that only about 60 percent had been attending meetings. MacFarquhar, *The Hundred Flowers*, 99.

45. On the question of nonparticipation in non-DPG group situations, see Martin King Whyte, *Small Groups and Political Rituals in China* (Berkeley, 1974), 45f.

46. Whether a chapter reports directly to the provincial-level organization or to an intermediate level often depends not on the structure of the DPG but on the level of the unit to which the chapter is attached. For example, if a chapter is in a university administered by the provincial government, it will report to the provincial DPG headquarters.

47. The DL, NCA, and RG each have fifty; others have smaller numbers. The DPG congress decides which people will go, but I have not determined how much autonomy these congresses have.

48. However, in other respects, DPG cadres appear to be more accountable to CP authorities than to the government bureaucracy. Such a cadre's dossier is kept by the CP's United Front Department (at least after he retires). *Chinese Law and Government* (Fall 1984), 127.

49. For a discussion of how the DPGs were financed before the Cultural Revolution, see van Slyke, *Enemies and Friends*, 217.

50. Xinhua, December 21, 1983, JPRS CPS-84-007, 76.

51. Xinhua, September 1, 1984, JPRS CPS-84-063, 54.

52. Radio Nanning, June 23, 1984, JPRS 84-050, 12.

53. Xinhua, November 5, 1983, JPRS 84844, 42-43. For additional information on DPG private schools, see Xinhua, December 5, 1985, JPRS CPS-86-009, 64f; *BJRB*, August 23, 1985, 1, JPRS CPS-86-002, 94f.; and Wang Xuexiao, "Zhejiang."

54. Between 1981 and 1986, thousands of DPG (and Federation of Industry and Commerce) people were engaged in 3,300 borderland development projects.

55. Liu Jinghuai, "Democratic Parties," 24.

56. Xinhua, October 29, 1983, JPRS 84857, 56. Between 1982 and 1984, 7,000 DPG people were elected to people's congresses at various levels. *Beijing Review*, January 16, 1984, 9.

57. Typically, prospects are invited to hear a lecture, in the hope that a few will be interested in joining the DPG.

58. Xinhua, November 4, 1983, JPRS 84857, 52

59. Li Rongxia, "Democratic Parties," 19.

60. On these publications, see chapter 5, note 44. Now *Guangming Daily* is directly run by the Communist Party and is aimed at intellectuals.

61. Li Honglin, "From 'One Person Alone Has the Say' to 'Everybody Has a Say,'" *Cun yan*, April 22, 1985, translated in FBIS, April 30, 1985, K-14ff. Li is president of the Fujian Academy of Social Sciences. On *Cun yan*, see Xinhua, April 20, 1985, JPRS CPS-85-051, 26; and *Far Eastern Economic Review*, June 20, 1985, 82f.

62. See Xinhua, April 20, 1985, JPRS CPS-85-051, 26, and *Far Eastern Economic Review*, June 20, 1985, 82f.

Chapter 8

1. In foreword to Carol Hamrin and Timothy Cheek, eds., *China's Establishment Intellectuals*, x.

2. The main work on this subject is Whyte, *Small Groups and Political Rituals in China*; it does not discuss the DPGs.

3. China's democratic activists tend to have disdain for the democratic parties. See James D. Seymour, *The Fifth Modernization: China's Human Rights Movement, 1978-1979* (Standardville, N.Y., 1980), 234.

4. This may seem a surprising assertion. However, during the early part of the Cultural Revolution (when the DPGs were inactive), China was probably more democratic (which is not to

say pleasant) than it is in 1986. At that time, ordinary people were fairly free to engage in politics--though ineffectively, because there was little law or institutional framework.

5. In 1983, ten of the twenty NPC Standing Committee vice-chairs were non-Communists, including seven DPG figures: Shi Liang (DL), Hu Yuzhi (DL), Hu Juewen (NCA), Rong Yiren (NCA), Su Deyan (Jiusan), Zhu Xuefan (RG), and Zhou Gucheng (PW).

6. In the future, as more people obtain higher education, the theoretical size could be large.

7. *PD*, August 29, 1957, *SCMP* 1606.

8. Nevertheless, the democratic parties do serve as significant bridges between China and foreigners (especially ethnic Chinese). They often host foreign delegations and encourage foreign investment in China. (See Li Rongxia, "Democratic Parties," 20.) The RG and Taiwan League, furthermore, are a potential line of communication with people on Taiwan.

9. See, for example, Xinhua, September 28, 1984, FBIS, October 2, 1984, in which the various DPGs embrace the Sino-British accord regarding Hong Kong.

10. Xinhua, February 1, 1986, JPRS CPS-86-023, 29.

11. This theme is developed in Joseph Fewsmith, *Party, State and Local Elites in Republican China: Merchant Organizations in Politics in Shanghai, 1980-1930* (Honolulu, 1985).

12. I will discuss this in a forthcoming book.

13. Julius Gould and William L. Kolb, *A Dictionary of the Social Sciences* (New York, 1964).

14. Quoted in Fewsmith, *Party, State and Local Elites*, 163.

15. Ibid., 190 (his italics).

16. The Association for Promoting Democracy has been paid scant attention in this book, but it is discussed in Gordon White, "Distributive Politics and Educational Development," especially 115f. The APD has been active training teachers for the poorer parts of the country, having trained 6,600 during 1984-85.

17. The baseline for this assertion (and for much of the following discussion) is provided by Andrew J. Nathan, "A Factionalism Model for CCP Politics," *China Quarterly* (January 1973), 37. My impression is that perquisites are dispensed on the basis of merit. The above statement regarding lack of clientelism would have to be modified to the extent that there is cronyism.

18. *Gandan xiang-zhao; rongru yu-gong.*

19. On recruitment of younger leaders in Shanghai, see dispatch of Shanghai City Service, March 10, 1986, JPRS CPS-86-028, 54.

20. Xinhua, November 22, 1985, JPRS CPS-85-121 gives the average age as "almost 75." Figures are for all eight groups, including the Zhi Gong Dang and Taiwan League.

21. Li Rongxia, "Democratic Parties," 19. Figure is as of February 1986 and includes the Zhi Gong Dang and Taiwan League.

22. One non-DPG scholar (a historian whom I know well and regard highly) was unable to introduce me to any democratic party members. He said none of this friends belonged, and he added: "Actually, we don't think very highly of the DPGs."

Bibliography

Chinese-language books and pamphlets for which no translation is available are given in pinyin; otherwise the titles have been translated. In practice, this usually means that where the name of the author is known, the title is translated, followed by the indication "(in Chinese)."

Alitto, Guy. *The Last Confucian: Liang Shu-ming and the Chinese Dilemma of Modernity.* Berkeley: University of California Press, 1979.
Boorman, Howard L., ed. *Biographical Dictionary of Republican China.* New York: Columbia University Press, 1967.
_____. *Men and Politics in Modern China (Preliminary).* New York, 1960.
Chang, Carsun (Zhang Zhunmai). *The Democratic Socialist Party's Platform Explained* (in Chinese). N.p., n.d. (Pamphlet, probably written around 1946-47).
_____. *Third Force in China.* New York, 1952.
Chao Ying. "Leftists of the Democratic Parties Viewed in the Light of the Rectification Movement." *Zhengming*, July 1958, *ECMM* 145.
Chen Mingshu. "My Self-Examination." *PD*, July 16, 1957, *CB* 470.
Chen Qing. "Refutation of Zhang Naiqi's 'My Self-Examination.'" Beijing *DGB*, August 10, 1957, *CB* 475.
Chen Qu. "Refutation of Zhu Anping's Preposterous Theory of 'Party Empire.'" *Xuexi*, August 3, 1957, *ECMM* 108.
Chen Xinkuei. Text of Confession. *GM*, June 20, 1957, *SCMP* 1563.
Chen Yi. "Report to the First National Construction Association Congress" (summary), Xinhua, April 13, 1955, *CB* 327.
Chen Zhuzhun, ed., *Studies in the History of the China National Construction Association* (in Chinese). Beijing, Chinese People's University Press, 1985.
Chen Ziyun. "On Multiparty Cooperation Under the Leadership of the Communist Party." *GM*, April 2, 1983, 3, FBIS,

April 17, 1984, K14.
Ch'en, Jerome, *Mao and the Chinese Revolution*. London: Oxford University Press, 1965.
_____, ed., *Mao*, Englewood Cliffs, N.J.: Prentice-Hall, 1969.
Cheng Peng. "An Injustice Hidden for Thirty Years: . . . The Story of the Execution of Zhou Yixiang," *Zhengming*, July 1983, 31-32, JPRS 84301, 102-105.
Chi Fang. "Report of the PW-CC" (summary), *GM*, December 21, 1958, *CB* 547.
Ch'ien Tuan-sheng. *The Government and Politics of China*. Cambridge: Harvard University Press, 1950, reprinted 1961.
Chu Anping. "Allow Me to Offer Some Opinions to Chairman Mao and Premier Zhou," *PD*, June 2, 1957, *SCMP* 1550.
_____. "Surrender to the People." *PD*, July 14, 1957, *CB* 470.
Cole, Allen B. "The United Front in the New China," *The Annals of the American Academy of Political and Social Science* (September 1951).
Deng Chumin. "Some Views on 'Deng Chumin Considers There is "Close" in "Bloom," ' " *GM*, April 26, 1957, *SCMP* 1541.
Fang Yirui. "The Democratic Parties, Past and Present" (in Chinese), *Baixing*, December 1, 1983, 53-55.
Fei Xiaotong. "I Admit My Guilt to the People," *PD*, July 14, 1957, *CB* 470.
Feng Hefa. "The Question of the Transformation of National Capitalists" (in Chinese). Beijing *DGB*, December 16, 1956.
Fisher, Tom. "Wu Han: The 'Upright Official' as a Model in the Humanities." In *China's Establishment Intellectuals*, ed. Carol Hamrin and Timothy Cheek. Armonk, N.Y.: M. E. Sharpe, 1986.
Forty Years of the China Democratic League: 1941-1981 (book in Chinese). Beijing: DL Central Historical Materials Committee: 1981.
Goldman, Merle. "The Unique 'Blooming and Contending' of 1961-62," *China Quarterly* 37 (January 1969).
Goodman, David S. G. *Groups and Politics in the People's Republic of China*. Armonk, N.Y.: M. E. Sharpe, 1984.
Gu Zhizhong. "Relations between the Communist Party and the Democratic Parties," *Xuexi*, December 12, 1956, *ECMM* 69.
Hamrin, Carol Lee, and Timothy Cheek, eds. *China's Establishment Intellectuals*. Armonk, N.Y.: M. E. Sharpe, 1986.

He Xiangning. "The Need for the Rectification of Working Style in the Revolutionary Guomindang," Xinhua, June 21, 1957, *SCMP* 1561.
Heping minzhu tongyi jianguo zhi lu (The road to peace, democracy, unification and reconstruction). Hong Kong: Minxian yuekan she, 1945. (Documents pertaining to the Democratic League, most of which also appear in *Minzhu tongmeng wenxian*.)
Hinton, Harold C. *Contradictions in Communist China: The Shattering of the Democratic Facade.* RAND Report, 1957.
_____. "The 'Democratic Parties': End of an Experiment," *Problems of Communism* (May-June 1958).
Hou Mingfang. "Surviving Bourgeois Rights Are Not a Bourgeois Right," *PD*, January 3, 1959, *SCMP* 1957.
Hu Quli, Speech Commemorating the Fortieth Anniversary of the Founding of the Association for Promoting Democracy. Xinhua, December 30, 1985, FBIS, January 3, 1986.
Hu Xiyuan. "Why Do We Adopt the Policy of Long-Term Coexistence and Mutual Supervision Between the Communist Party and the Democratic Parties?" *Shishi shouci*, August 10, 1956, *ECMM* 54.
Hu Yihe. "Empty Talk Can Do No Good in the Question of Relations Between the CCP and the Democratic Parties at the Primary Level," *GM*, January 17, 1957, *SCMP* 1463.
Hu Yuzhi and Li Wenyi. "Cherishing the Memory of Comrade Wu Han: Straightforward Talk from a Straightforward Person, Utter Devotion to Friends" (in Chinese), *PD*, October 3, 1979, 3.
Huang Fenyi, "Footprints Forty Years Before: Cherishing the Memory of My Father Huang Yanpei" (in Chinese), *Beimei ribao* (New York), March 25, 1986, 4.
Huang Jixiang. "A Request for the People's Forgiveness," *PD*, July 14, 1957, *CB* 470.
_____. "Two Problems in the Work of the Democratic Parties," *GM*, January 3, 1957. *SCMP* 1457.
Huang Shaohung. "A Review of My Mistakes and Crimes," *PD*, July 16, 1957, *CB* 470.
Huang Yanpei. Introductory Speech before the NCA Central Committee (in Chinese), *Xinhua ban yue kan*. Reprinted from *GM*, November 6, 1956.
_____. *The Past Eighty Years* (in Chinese). Beijing: Wenshi ziliao chuban she, 1982.

_____. "To Strengthen the Unity, and Work Hard Under the Leadership of the CCP," Xinhua, April 1, 1955, *SCMP* 1023 (also *CB* 327).
Jing Xu. *The Multi-Party System under Communism* (in Chinese). Hong Kong: Freedom Front, 1956.
Kennedy, Melville T. "The Chinese Democratic League," *Papers on China from the [Harvard] Regional Studies Seminars*, vol. 7. Cambridge, Mass., 1955.
Kertesz, Stephen D., ed. *The Fate of East Central Europe: Hopes and Failures of American Foreign Policy*. Notre Dame, Ind., 1956.
Klein, Donald W., and Anne B. Clark. *Biographic Dictionary of Chinese Communism, 1921-1965*. Cambridge: Harvard University Press, 1971.
Li Boqiu. "Responsibilities and Work of the PW" (in Chinese), *GM*, June 5, 1956.
Li Jishen. Address Dedicated to Mao Zedong, Delivered on the Occasion of the Tenth Anniversary of the People's Republic, Xinhua, September 28, 1959, *CB*, 594.
_____. Address to the Third Plenum of the Second Revolutionary Committee Central Committee. *GM*, January 10, 19953, *SCMP* 503.
_____. "My Party and What It Stands For," *People's China*, November 16, 1956.
_____. "Report to the Central Committee of the Revolutionary Committee" (summary). *GM*, December 16, 1958, *CB* 547.
Li Rongxia. "Democratic Parties Work for Modern China," *Beijing Review*, February 24, 1986.
Li Weihan. "The Democratic Parties Must Carry Out Their Basic Self-Transformation," *GM*, Nov. 6, 1957, *SCMP* 1663.
_____. "The Democratic United Front in China," Xinhua, June 25, 1956, *CB* 402.
_____. "Further Strengthening the United Front Work Within the Government," Xinhua, April 29, 1951, *CB* 96.
_____. "Our People's Democratic United Front," *People's China*, August 16, 1956.
_____. "The United Front Work and the Party," Xinhua, September 25, 1956, *CB* 418.
Li Zhuchen. Work of the NCA Central Committee at the Second Session of Its First Congress--Extracts (in Chinese). *Xinhua ban yue kan*, December 21, 1956.

Lin Yaohua. "The Treacherous and Ugly Fei Xiaotong," *PD*, August 2, 1957, *CB* 475.
Liu Dongyan. *My View of the China Youth Party's Past, Present and Future* (in Chinese). N.p, 1963.
Liu Jinghuai, "The Democratic Parties Are a Vital Force in the Four Modernizations Drive: A Visit with Li Ding, Deputy Director of the United Front Work Department," *Liaowang*, April 1, 1985, 12-13, JPRS CPS-85-063, 24-27.
Liu Shaoqi. Political Report to the Eighth CCP Congress. September 16, 1956, *CB* 412.
Liu Xia, comp. *Eighty Years of the China Youth Party* (in Chinese). Chengdu: Guo hun shudian, 1941.
Liu Xiaoping and Xu Shuang. "Characteristics and Merits Unique to China's Political System," *GM*, May 7, 1984, JPRS CPS-84-041, 14.
Long Yun. "My Ideological Review," *PD*, July 14, 1957, *CB* 470.
Luo Longji. "Band the Non-Party Intellectuals Closer with the Party," *PD*, March 23, 1957, *CB* 444.
_____. "My Preliminary Examination," *PD*, July 16, 1957, *CB* 470.
_____. "The Question of the Higher Intellectuals," Xinhua, June 26, 1956, *CB* 402.
Lu Dingyi. "Let All Flowers Bloom Together, Let Diverse Schools of Thought Contend," *PD*, June 13, 1956, *CB* 406.
Lu Yi'an, *The Split in the Democratic Socialist Party* (in Chinese). N.p., n.d. (Pamphlet, apparently written in the late 1940s.)
Ma Ding. "Our Province Has Implemented United Front Policy for Many People," *Sichuan Daily*, February 8, 1985, 1, JPRS CPS-85-040, 89-90.
MacFarquhar, Roderick. *The Hundred Flowers*. London: Atlantic Books, 1960.
_____. *The Origins of the Cultural Revolution*. New York: Columbia University Press, 1974.
Major Documents of the Guomindang Revolutionary Committee (series of books in Chinese). Beijing: RG Central Propaganda Department, vol. 1, 1959; vol. 2, 1982.
Mao Tse-tung [Mao Zedong]. *On the Correct Handling of Contradictions Among the People* (Beijing, 1957). (Note: this is the official version as published a few months after the speech was delivered. The original was significantly differ-

ent. Extracts of the original appeared in the *New York Times*, June 13, 1957, and were reprinted in MacFarquhar, *The Hundred Flowers*, 265-77.)
_____. *Selected Works of Mao Tse-tung*. Beijing: Foreign Languages Press, 1964-77. 5 vols.
Min Ganghou. "An Emergency Conference Convened by Zhang Bojun," *GM*, July 4, 19576, *SCMP* 1571.
Minzhu tongmeng erh zhong quan hui zhengzhi baogao (Political Report to the Second Plenum of the Central Executive Committee of the Democratic League). Shanghai: Zhongguo minzhu tongmeng zongbu, 1947.
Minzhu tongmen wenxian (Documents on the Democratic League). N.p., Zhongguo minzhu tongmeng zongbu, 1946.
Ou Zhibei. "Difficulties in Organizational Work of the Democratic League in Shaanxi and Gansu," *GM*, February 7, 1957, *SCMP* 1476.
Pauw, Alan. "Chinese Democratic Parties as Mass Organizations." Masters thesis, 1981. (Citations refer to the above. A version was published in *Asian Affairs* [July/August 1981], 372-90, under the title "Chinese Democratic Parties as a Mass Organization.")
Ping Xin. *On the Third Force and the Democratic Movement* (in Chinese). Hong Kong: Zhihshi chuban she, 1947. (Pamphlet arguing the case for a political force independent of the Communists and Nationalists.)
"Red China's Eight 'Democratic Parties': New 'United Front' Tool," *Inside China Mainland* (Taipei) (March 1984).
Renmin shouci (People's Handbook). (The 1955 edition contains brief sketches of the activities of the minor parties.)
Sa Gongliao. "The China Democratic League," *People's China*, April 1, 1957.
Seymour, James. D. "Communist China's Bourgeois-Democratic Parties." Master's essay, Columbia University, 1960.
Shen Junru. "Work Report of the Second Central Committee of the Democratic League" (summary), *GM*, December 18, 1958, *CB* 547.
Shen Zhiyuan."Long-Term Coexistence and Mutual Supervision of Parties," *People's China*, March 16, 1957.
Shi Liang. "Speech at the Fortieth Anniversary Meeting of the China Democratic League" (pamphlet in Chinese). Beijing: DL Central Committee, 1982.
Shi Qi and Sun Nan. "How to Understand the Policy of Long-

Term Coexistence Between the Communist Party and the Democratic Parties," *Zhengzhi xuexi*, September 13, 1956, *ECMM* 56.
Shi Rugang."My Understanding of 'Long-Term Coexistence and Mutual Supervision,' " *Xuexi*, February 3, 1957, *ECMM* 78.
Stein, Gunther. *The Challenge of Red China*. New York, 1945.
Stuart, John Leighton. *Fifty Years in China*. New York, 1954.
Sun Xiaocun, Feng Hofa, and Zhang Fan. *A History of the China National Construction Association* (in Chinese). Beijing: National Construction Association and National Federation of Industry and Commerce, 1983.
Tan Diwu."Why Have I Committed Such Serious Crimes?" *PD*, July 15, 1957, *CB* 470.
Ting Yang. "Why Bourgeois Intellectuals Are Said to Be Part of the Exploiting Class," *Wen hui bao*, June 24, 1958, *SCMP* 1814.
Tong Yi. "The Democratic Parties in Action," *Beijing Review*, November 7, 1983, 22-28.
van Slyke, Lyman P. *Enemies and Friends: The United Front in Chinese Communist History*. Stanford: Stanford University Press, 1967.
Wang Jingwei et al. *Jihui zhuyi de disan dang* (The opportunist Third Party). Hebei: Zhongguo guomindang Hebei sheng dangwu zhidao weiyuanhui xuanquan bu, 1928. (In addition to Wang and various pseudonymous writers, this book includes articles by Chen Gongbo, Lin Bosheng, Shi Cuntong, Huang Hanrui--all highly critical of the Third Party.)
Wang Shaoao. "Report of the PW CC" (summary), *GM*, December 21, 1958, *CB* 547.
Wang Xuexiao, "Zhejiang Democratic Parties Promote Education," *PD*, September 20, 1985, 3, JPRS CPS-85-117, 34f.
Wei Xiutang. "China's Other Parties Today," *China Reconstructs* (April 1984), 28-31.
Wen Yiwen. "'Pernicious Expansion' Has Corrupted the Jiusan Society," *GM*, August 29, 1957, *SCMP* 1623.
White, Gordon. "Distributive Politics and Educational Development: Teachers as a Political Interest Group." In *Groups and Politics in the People's Republic of China*, ed. David S. G. Goodman. Armonk, N.Y.: M. E. Sharpe, 1984.
Wu Dakun. "Zhang Naiqi as I Know Him--Political Adven-

turer of the Bourgeois Class," Xinhua, July 3, 1957, *SCMP* 1570.

Wu Han. "How to Unite Activities of the China Democratic League with Activities of Institutions of Higher Education," *GM*, April 20, 1953, *SCMP* 580.

———. "Self-Criticism on 'Dismissal of Hai Rui,' " *PD*, December 30, 1965, *CB*, 783.

Xi Zhongxun. Speech Commemorating the Fortieth Anniversary of the Jiusan. Xinhua, September 2, 1985, JPRS CPS-85-103, 22-25.

Xiao Yena and Wang Erde. *Communist China's Democratic Parties* (in Chinese). Hong Kong, 1951.

Xu Changtai. "Hastening the Transformation of the Democratic Parties into Political Forces That Would Truly Serve Socialism," *GM*, April 20, 1958, *SCMP* 1765.

Xu Zhengfan. "In Refutation of the Argument That 'The Class Struggle Has Ended,' " *PD*, November 6, 1957, *SCMP* 1474.

Xu Baoju. "The Ideological Education Work of the Democratic Parties," *GM*, February 5, 1957, *SCMP* 1474.

Xu Daohe. " 'Mutual Supervision' or 'Stand on Equal Terms?' " *Xuexi*, August 3, 1957, *ECMM* 105.

Xue Yu. "My New Birth," *PD*, June 8, 1959, *CB* 583.

Ya Qiao. "Sun Xiaocun," *Xingdao wanbao*, November 6, 1951, *SCMP* 213.

Yan Qi and Wang Youqiao, eds. *Studies in the History of the China Peasants' and Workers' Democratic Party* (in Chinese). Beijing: Chinese People's University Press, 1984.

Yan Xinming. "Listen to Zhang Bojun's Delirious Statement," *PD*, July 3, 1957, *SCMP* 1571.

———. "Zhang Bojun Is Determined to Rebel," *PD*, August 4, 1957, *CB* 475.

Yang Hanhui. *Political Education in Contemporary China* (in Chinese). Beiping: Renwu shudian, 1932. Contains a chapter on the minor parties, primarily the Youth Party, Third Party, Agricultural Socialist Group, and the Democratic Socialists.

Yao Wenyuan, "On the New Historical Play 'Dismissal of Hai Rui,' " Shanghai *WHB*, reprinted in *PD*, November 30, 1965, *CB* 783.

You Guangyuan. "The Class Nature of the Present People's Democratic Dictatorship in China," *Xuexi*, November 2,

1956, *ECMM* 66.

Yu Gang and Yu Jianze. "A Brief Discussion on the Historical Path of Chinese Democratic Parties--Understanding Gained from Studying *Selected Works of Deng Xiaoping*," *Red Flag* 23 (December 1983), 17-22, JPRS CRF-84-002.

Yu Hai. "Le rôle de la bourgeoisie nationale dans la révolution chinoise," *Cahiers du communisme* (August 1950).

Yu Runtang and Yao Chuankeng. *Contemporary Political Parties in China* (in Chinese). Guangzhou: Zongheng wenhua shiye gongxi, 1948. (Discusses many groups, including some tiny ones overlooked in other works.)

Zhang Bojun. "Bring the Democratic Parties into Full Play," *PD*, March 19, 1957, *CB* 444.

_____. "The Democratic League's Participation in the Cultural-Educational Construction of the State," *PD*, July 25, 1953, *SCMP* 644.

_____. "I Bow My Head and Admit My Guilt before the People," *PD*, July 16, 1957, *CB* 470.

_____. "I Have Committed a Serious Mistake Politically," *PD*, June 14, 1957, *SCMP* 1558.

_____. "I Thoroughly Remold Myself and Become a New Man," *PD*, May 4, 1959, *CB* 583.

_____. "The Road to Socialism Must Be Followed," *GM*, June 12, 1957, *SCMP* 1561.

Zhang Guofan. "Why Do We Say That the Leadership of the Communist Party Is Absolute?," *Zhengming*, September 10, 1958, *ECMM* 149.

Zhang Naiqi. "Lay Down the Arms of Old Democracy and Surrender to New Democracy," Shanghai *DGB*, March 2, 1952, *SCMP* 289.

_____. "My Self-Examination," *PD*, July 16, 1957; *CB*, 470.

_____. "Several Questions Concerning the Work of Assisting and Guiding the Transformation of Industry and Commerce," Beijing *DGB*, June 9, 1957, *SCMP* 1570.

_____. "Work Report of the China National Construction Association" (summary), Xinhua, April 14, 1955, *CB* 327.

Zhang Youren. "Revival of Bourgeois Economics Cannot Be Tolerated," *Jianshe*, September 3, 1957, *ECMM* 108.

Zhang Zhiyi, comp. *Political Parties and Groups During the Anti-[Japanese] War* (in Chinese). N.p., 1939. (Has useful information on the early histories of the Third Party, National Socialist Party, National Salvation Association, and other groups.)

———. "On the Recognition of the Policy of Long-Term Coexistence and Mutual Supervision," *Xuexi*, January 18, 1958, *ECMM* 124.

———. "The People's Democratic United Front Is a Special Form of Class Struggle," *GM*, July 20, 1958, *SCMP* 1827.

———. "Political Study Is an Important Means of Helping Bourgeois Elements and Democratic Groups Remold Themselves," *GM*, January 1, 1959; *CB*, 547.

———. "Problems Concerning the People's Democratic United Front," Beijing *DGB*, March 31, 1957, *CB*, 444.

Zhang Zhizhong. "Forge Closer Ties Between the Communist Party and Non-Party Circles," *PD*, March 9, 1957; *CB* 444.

Zhongguo minzhu tongmeng san zhong quan hui (The Third Plenum of the Central Executive Committee of the China Democratic League). N.p. Compiled by the headquarters of the DL, 1948. (Pamphlet containing three resolutions adopted in January 1948.)

Zhongguo qingnian dang shih lue ji zhengkang (Outline of the history and platform of the China Youth Party). N.p.: Zhongguo qingnian dang Guangdong sheng dangbu, 1947. (Pamphlet, probably published in Guangzhou.)

Zhongguo xin minzhu yundong zhong de dangpai (Political parties and groups in China's New Democracy movement). Shanghai: Xin Zhongguo chuban she, 1946. (Reprints, some dealing with the DPGs.)

Zhongguo ge xiao dangpai xiankuang (The situation of China's small parties). N.p., 1946.

Zhong Xun."The Supervision of Friends," *PD*, January 15, 1957, *SCMP* 1463.

Zhou Enlai. "On the Question of Intellectuals," Xinhua, January 29, 1956, *CB* 376.

Zhou Fuzheng. "Create a New Style in Study," *Zhengming*, June 1957, *ECMM* 98.

Zhou Jingwen, *Fengbao shi nian,*. Hong Kong: Shidai piping Press, 1959. (Available in English as Chou Ching-wen, *Ten Years of Storm: The True Story of the Communist Regime in China.* New York: Holt, 1960.)

Zhu Juyang. *How Do the Chinese Communists Handle the Democratic Parties?* (in Chinese). Hong Kong, 1952.

Index

Academy of Sciences, 91
Admissions standards of DPGs, 72. See also Constituency under name of DPG
Age of DPG people, 72, 83, 91, 134n30
Army, 73, 115n8
Association for Promoting Democracy, 58, 90, 119n29, 133n16; background of, 21; constituency, viii, 28, 43, 129n25; individuals in, 71, 123n63, 127n12; size of, viii

Bourgeoisie, 54, 61; DPGs as organizations for, 58; Mao on, 10; mentality, 39f., 60; petty, 8; reformation of, 36, 40-42. See also Intellectuals
Business people, 34, 36, 38, 60, 82, 122n39, 122n42

Cadres (DPG), 79f., 89
Cai Tingkai, 19, 21, 52
Cao Yi'ou, 66
Campaigns involving DPGs, 116n29. See also name of individual campaigns
Canton Commune, 19
Central Socialist Academy, 71
Chang, Carsun (Zhang Zhunmai), 13
Chapters. See Local DPG affairs
Chen Mingshu, 19
Chen Qitong, 61
Chen Shiwei, 52
Chiang Kai-shek, 14, 18f.
China Revolutionary Party, 111n3
Chinese Academy of Sciences, 91
Chinese People's Political Consultative Conference, 23, 30f., 37, 41, 52f., 76f., 95, 98, 118n12, 130n33, 35, 131n47
Chu Anping, 54f., 56, 60, 122n46
Chu Tunan, 70

Chu Yuxian, 123n63
Classes (social), theory of, 7
Colleges. See Education
Cominform (1947), 11
Comintern (1920), 6
Common Program (1949), 23
Communist Party: DPG relations with, 33, 38, 42f., 45f., 49, 55, 57, 68f., 71f., 79f., 83f., 87f., 91, 96f., 99, 115n12, 117n51, 122n35, 124n87, 131n41, 42; rank-and-file relations with DPGs, 24, 41, 70. See also Cadres (DPG)
Congress. See National People's Congress
Constituencies of DPGs, viii. See also individual DPGs
Constitutions of DPGs, 70; text (RG), 98-108
Constitutions of PRC, 30, 116n36
Consulting work (for enterprises), 82, 84
"Contradictions," 48
Corporate state (political theory), 88
Counterrevolutionaries. See Opposition to communism
Cultural Revolution, 64-67, 74, 76f.
Cun yan, 84
Czechoslovakia: multiparty system, 4f

Dai Li, 17
Democracy movement (1978-1981), 69
Democratic Constitutionalist Party, 13
Democratic League, 51, 55, 57f., 60, 64, 67, 68, 75f., 81-83, 97, 131n41; background of, 10, 14-18, 79, 113n23, 114n36; campaigns, 34; constituency, viii, 28f., 79, 84; dismissals, 73; emergency conference (1957),

145

146 China's Satellite Parties

95-97; individuals in, 32, 38, 52, 59, 61, 65f., 70, 133n5; organizational changes, 42; relations with other DPGs, 20; size, viii, 42-45, 53, 118n13, 119n34, 124n76
Democratic parties and groups (term discussed), vi, 25, 29, 86f.
Democratic Scientific Forum (Society), 20
Democratic Socialist Party, 13, 112n11
Deng Xiaoping, 61, 64, 69, 86
Deng Yanda, 13
Dictatorship of the proletariat, 11
Discipline, 102f.
Dismissals, 73
Dues, 81
Duties of DPG members, 101

Economy. See Modernization
Education 52, 84, 96, 133n16; DL academic policies, 93f; institutions run by DPGs, 81f.
Elections: government, 32, 83, 96; within DPGs, 79-80
Ethnic minorities, 27, 72, 128n24, 129n27

Farmers, 73
Federation of Industry and Commerce, 72, 129n27
Fei Xiaotong, 95f., 130n35
Fewsmith, Joseph, 89
Finances of DPGs, 80-81, 113n23
Five-anti campaign, 34, 117n33
Foreign relations, 30, 133n8
Four Basic Principles, 84
Four Modernizations. See Modernization
Free speech, 80. See also Hundred Flowers

Germany, East: multiparty system, 4
Government, DPG people in, 30-34, 41, 51, 60, 66f., 119n4
Grand League of Democratic Political Groups, 14
Great Leap Forward, 63
Groups. See Interests (political); Small groups
Guangming Daily, 52f., 54f. 57, 84
Guomindang, 6, 89f., 116n17; CP relations with, 9f., 15, 22; DPG relations with, 12, 13, 15, 18, 20, 57, 59, 74, 98, 100, 111n10
Guomindang Revolutionary Committee. See Revolutionary Guomindang
Guo Muoro, 64

"Hai Rui Dismissed from Office," 65, 68
He Xiangning, 128n25
Hong Kong, 14, 18-20, 69, 100, 111n9
Hu Juewen, 128n22, 133n5
Hu Yaobang, 70
Hu Yuzhi, 59, 70, 95, 133n5
Hu Zi'ang, 51
Huang Jixiang, 44
Huang Yanpei, 13, 20, 28, 64
Huang Yaomian, 95
Hundred Flowers, 33, 46-63, 66, 71, 78, 85f., 119n3, 120n15
Hungary, 96

"Immortals," meetings of, 63
Industry and Commerce, Federation of, 72, 132n54
Intellectuals, 3, 53, 60, 77, 85, 96, 117n46; DPGs as parties for, 27f, 43, 88; Mao Zedong on, 8, 48f.; reform of, 34, 38; Zhou Enlai on, 36f.
Interests (political), 29, 89
International relations, 30, 133n8
Interpellation, 77
Israel, John, 85

Ji Fang, 128n22
Jin Ruonian, 95
Jiu-san Study Society, 52, 58, 78, 81, 88; background of, 20f.; constituency, viii, 72, 84, 128n25; individuals in, 116n36, 128n22, 133n44, 45; organizational changes, 42; size, viii, 42, 44

Kang Sheng, 64, 66
Korean War, 27, 34, 40, 58
Kuomintang. See Guomindang

Legal system, 51
Legitimization, 87
Lei Zhen, 112n19
Lenin, Vladimir, 6, 89
Li Chai-sum. See Li Jishen

Li Honglin, 84
Li Jishen, 19f., 57
Li Junlong, 27, 115n8
Li Renren, 51
Li Weihan, 25-27, 33, 37, 40f., 43, 60, 66
Li Xiyuan, 111n4
Li Zhuchen, 119n4
Li Zongren, 116n17
Liang Shuming, 13
Liu Binyan, 84
Liu Jiaxi, 91
Liu Shaoqi, 30f., 38, 40, 61, 64, 112n20, 117n49
Local DPG affairs, 43ff., 52, 57, 59, 64, 67, 78f., 80, 91, 97, 106-108, 114n2, 119n29, 131n44
Long Yong, 43, 57f., 71, 123n62
Longevity of DPGs, 39-41
"Long-term coexistence," 39-42, 57f., 59, 100
Lu Dingyi, 47, 40
Lu Xun, 128n22
Luo Longji, 52f., 59, 60, 62, 66, 119n4, 124n88; background of, 17, 112n19, 113n23; and DL expansion, 42f., 45

Ma Bi, 116n17, 127n12
Ma Hongbin, 129n27
Ma Xulun, 21
Mai Gongbin, 126n4
Mao Zedong, 6, 86; criticism of, 62, 121n31; "On Coalition Government," 10; "On Contradiction," 119n7; on intellectuals, 47; on longevity of parties, 40; on a multiparty system and mutual supervision, 46, 49; on social classes and united fronts, 7, 22; "On the Correct Handling of Contradictions among the People," 48-49, 120n13; "On the New Democracy," 8f.; protects DPGs during Cultural Revolution, 66
Marshall, George, 15, 21-22, 112n20
May Forums (1957), 50, 54, 121n16, 121n17
May Fourth Movement (1919), 97
Mei Gongbin, 67
Membership: application, 102. See also Sizes of DPGs; Constituency under name of DPG
Mensheviks, 3

Military, 73, 115n8
Min Ganghou, 95
Min-ge, 26. See also Revolutionary Guomindang
Min-jian. See National Construction Association
Min-meng. See Democratic League
Minorities, 27, 72, 128n24, 129n27
Modernization: DPG contribution to, 77, 81-83, 99; DPGs as modern institutions, 90f.
"Mutual supervision," 39-42, 49, 58f., 100

Nanchang Uprising, 18f.
National bourgeoisie. See Business people
National Construction Association, 60, 67, 81f.; background of, 20; campaigns, 34; constituency, viii, 28f., 72, 83, 129n25; dismissals, 73; individuals in, 51, 54, 57, 64, 79, 128n22, 133n5; reform of members, 36, 115n14; size, viii, 42f., 45
National Liberation Action Committee, 111n3
National People's Congress, 31f.
National Salvation Association, 13f.
National Social Party, 13, 17, 112n11
Nationalist Party. See Guomindang

Opinion, 84
Opposition to communism, 43, 47, 49, 73, 78; counterrevolutionaries, 62, 64, 68, 121n31, 124n77. See also Hundred Flowers
Organizational features of DPGs, 25-30, 42-45, 79, 103-108

Peasants, 73
Peasants' and Workers Democratic Party, 51, 56, 58, 59; background of, 13, 21; constituency viii, 28, 43; individuals in, 52, 133n5; size, viii, 44, 53
Peng Dehuai, 65
Peng Zhen, 61, 64
People's political consultative conferences (PPCC), 31, 52, 77, 83, 130n33; defined, xi
Poland, 96; multiparty system, 4, 53
Political Consultative Conference

148 China's Satellite Parties

(pre-1949), 16
Privileges and perquisites of DPG membership, 67, 75f.
Programs of DPGs before 1949, 114n36
Provinces. See Local DPG affairs
Provisional Action Committee of the Guomindang, 111n3
Public Interest Party (Zhi Dong Dang), viii, xix, 116n18
Publications of DPGs, 122n44. See also name of publication

Qian Weichang, 95f.
Qian Zhangshao, 51

Red Guards, 64, 66f.
Rehabilitations, 68
Resignations from DPGs, 78f., 102
Revolutionary Guomindang, 52, 55, 58, 60, 75, 78f., 81f., 127n10; background of, 18-20, 67; constituency, viii, 27, 72, 100f., 129n27; constitution (text), 98-108, 129n27; elections, 32; individuals in, 51f., 57, 66, 71, 74, 115n6, 116n28, 126n4, 127n12, 128n25, 133n5; organization, 26f.; reform of members, 35; size, viii, 42, 44f.
Rightists, 57f., 63, 73f., 115n6
Rights of DPG members, 101f.
Rong Yiren, 133n5.
Rural Reconstruction Association, 13

Sa Gongliao, 117n51
San-min-zhu-yi Comrades Association, 19, 26
San-min-zhu-yi Promotion Association, 19, 21, 26
Sanxia, 78, 88
Schools. See Education
Schmitter, Philippe, 89
September Third Study Society. See Jiusan Study Society
Service, Richard, 20
Shaanxi-Gansu-Ningxia border region, 9
Shao Hengqiu, 129n27, 30
Shao Lizi, 51f.
Shi Liang, 32f., 66, 70, 95, 111n7, 128n19, 25, 133n5.
Sizes of DPGs, viii, 42-45, 91, 118n13. See also individual DPGs
Small groups, 85
Socialism, implications of, 39, 48
Song Qingling, 19, 128n25
Soong Ching-ling, See Song Qingling
Soviet Union, 16, 37, 98; DPG members' views of, 57, 123n63; multiparty system in, 3; view of China's DPGs, 11f., 110n20
Speech, freedom of. See Free speech; Hundred Flowers
Spiritual pollution, 69, 126n7
Stuart, John Leighton, 112n20
Students, 95-97
Su Deyan, 133n5.
Sun Yat-sen, 98f.

Taiwan, 30, 69, 98f., 100, 112n19, 127n10
Taiwan Democratic Self-Government League, viii, 115n16, 127n10, 128n25, 130n40
Tan Pingshan, 19
Tao Darong, 95
Teachers. See Association for Promoting Democracy; Education
Third Party, 13
Three-anti campaign, 34, 117n33
Three Gorges. See Sanxia
Three-thirds system, 7
Three People's Principles, 98f.
Transferring membership, 102
Tribune, 84
Truman, Harry S., 112n20

United Front: pre-1949, 6-11, 23, 110n19; post-1949, 29f. 34, 40, 47, 50, 52, 61, 98
United Front Department, 25f., 34, 56, 69, 80, 89
United National Construction League, 14
United States: relations with DPGs (pre-1949), 17, 19f., 21-22, 113n20
Unity, 55, 60
Universities. See Education

Violence against DPG people, 34
Vocational Educational Group, 13, 20

Wang Kunlun, 67, 71
Wang Zhixiang, 116n36

Wei Lihuang, 116n17
Wen Yiduo, 14
Women, 38, 72, 90, 115n3, 128n19
Workers, 44, 73, 111n10
Wu Dakun, 59
Wu Han, 125; attacks on, 65f., 66f.; dissent in early 1960s, 63; rehabilitated, 68; rightists attacked by, 59, 61-62, 64f.
Wu Qingchao, 95
Wu Wenzao, 127n12

Xie Xuehong, 128n25
Xu Deheng, 42, 58, 119n4, 128n22

Yan'an period, 10
Yang Jingren, 69
Yang Qing, 51
Yao Wenyuan, 65
Ye Duyi, 95
Ye Jianying, 68

Youth League, 73f., 96
Youth Party, 13, 16f., 112n11

Zeng Zhaolun, 95f.
Zhang Bojun, 52-54, 56, 60, 62, 71, 95, 97; and expansion of DL, 42-45, 58, 97; and PW, 13
Zhang Naiqi, 50, 54, 57, 59f., 122n39
Zhang Lan, 14f. 18, 114n36
Zhang Zhizhong, 52, 127n12
Zhang Zhunmai. See Chang, Carsun
Zhdanov, Andrei, 11
Zhou Enlai, 28, 33, 36f., 46, 66, 115n11, 117n46
Zhou Gucheng, 133n5.
Zhou Jianren, 71
Zhou Xinmin, 16
Zhou Yixiang, 116n28, 126n4
Zhu Xuefan, 133n5
Zhukov Georgy, 11

About the Author

A graduate of Yale University, James D. Seymour received a Ph.D. in Public Law and Government from Columbia University. He has taught at N.Y.U., the New School for Social Research, and Columbia and is currently a Visiting Research Fellow at Columbia's East Asian Institute.

Professor Seymour is the author of *China: The Politics of Revolutionary Reintegration* (1976), *The Fifth Modernization: China's Human Rights Movement, 1978-1979* (1980), and *China Rights Annals 1* (1985), the co-author of *Introduction to Comparative Politics* (1984), and the editor of *SPEARhead: Bulletin of the Society for the Protection of East Asians' Human Rights*.

East Gate Books

Harold R. Isaacs
RE-ENCOUNTERS IN CHINA

James D. Seymour
CHINA RIGHTS ANNALS 1

Thomas E. Stolper
CHINA, TAIWAN, AND THE OFFSHORE ISLANDS

William L. Parish, ed.
CHINESE RURAL DEVELOPMENT
The Great Transformation

Anita Chan, Stanley Rosen, and Jonathan Unger, eds.
ON SOCIALIST DEMOCRACY AND THE CHINESE LEGAL SYSTEM
The Li Yizhe Debates

Michael S. Duke, ed.
CONTEMPORARY CHINESE LITERATURE
An Anthology of Post-Mao Fiction and Poetry

Michiko N. Wilson
THE MARGINAL WORLD OF ŌE KENZABURO
A Study in Themes and Techniques

Thomas B. Gold
STATE AND SOCIETY IN THE TAIWAN MIRACLE

Carol Lee Hamrin and Timothy Cheek, eds.
CHINA'S ESTABLISHMENT INTELLECTUALS

John P. Burns and Stanley Rosen, eds.
POLICY CONFLICTS IN POST-MAO CHINA
A Documentary Survey, with Analysis

Victor D. Lippit
THE ECONOMIC DEVELOPMENT OF CHINA

James D. Seymour
CHINA'S SATELLITE PARTIES

Augsburg College
George Sverdrup Library
Minneapolis, Minnesota 55454